Lecture Notes in Mathematics

Edited by A. Dold and B. Eckmann

T0215402

426

Martin L. Silverstein

Symmetric Markov Processes

Springer-Verlag
Berlin · Heidelberg · New York 1974

Prof. Martin L. Silverstein
University of Southern California
Dept. of Mathematics
University Park
Los Angeles, CA 90007/USA

Library of Congress Cataloging in Publication Data

Silverstein, Martin L 1939-
Symmetric Markov processes.

(Lecture notes in mathematics ; 426)
Bibliography: p.
1. Markov processes. 2. Potential, Theory of.
I. Title. II. Series. Lecture notes in mathematics
(Berlin) ; 426.
QA3.L28 no. 426 [QA274.7] 510'.8s [519.2'33] 74-22376

AMS Subject Classifications (1970): 60J25, 60J45, 60J50

ISBN 3-540-07012-5 Springer-Verlag Berlin · Heidelberg · New York
ISBN 0-387-07012-5 Springer-Verlag New York · Heidelberg · Berlin

Offsetdruck: Julius Beltz, Hemsbach/Bergstr.

DEDICATED TO

W. FELLER

Introduction

This monograph is concerned with symmetric Markov processes and especially with Dirichlet spaces as a tool for analyzing them. The volume as a whole focuses on the problem of classifying the symmetric submarkovian semigroups which dominate a given one.

The main results are contained in Chapter III and especially in Section 20. A modified reflected space is determined by a boundary Δ together with an intensity for jumping to Δ rather than to the dead point. Every dominating semigroup which is actually an extension is subordinate to at least one modified reflected space. The extensions subordinate to a given modified reflected space are classified by certain Dirichlet spaces which live on the appropriate Δ. When the intensity for jumping to Δ vanishes identically, the subordinate extensions all have the same local generator as the given one. The most general dominating semigroup which is not an extension is obtained by first suppressing jumps to the dead point and/or replacing them by jumps within the state space and then taking an extension.

Some general theory is developed in Chapter I. A decomposition of the Dirichlet form into "killing", "jumping" and "diffusion" is accomplished in Chapter II. Examples are discussed in Chapter IV.

Each chapter is prefaced by a short summary.

The main prerequisite is familiarity with the theory of martingales as developed by P. A. Meyer and his school. Little is needed from the theory of Markov processes as such, except from the point of view of motivation.

For a treatment of classification theory in the context of diffusions we refer to [20] and [30]. In fact it is M. Fukushima's paper [20] that inspired our own research in this area and his influence is apparent throughout the volume.

For more information on "sample space constructions" for extensions of a given process we refer to Freedman's book [52] where current references to the literature can be found.

The expert typing was done by Elsie E. Walker at the University of Southern California.

Notations

Throughout the volume $\underline{\underline{X}}$ is a separable locally compact Hausdorff space and dx is a Radon measure on $\underline{\underline{X}}$ which charges every nonempty open set. The indicator of a set will be denoted both by 1_A and $I(A)$. The integral of a function ξ over the set determined by a condition such as "$X_t \in \Gamma$" will be denoted both by $\mathscr{E}[X_t \in \Gamma : \xi]$ and $\mathscr{E}\, I(X_t \in \Gamma)\, \xi$. The measure which is absolutely continuous with respect to a given measure μ and has density φ will often be represented $\varphi \cdot \mu$. The subcollection of bounded functions in $\underline{\underline{F}}$ will be denoted by $\underline{\underline{F}}_b$. All functions are real valued. In particular $L^2(dx)$ or $L^2(\underline{\underline{X}}, dx)$ is the real Hilbert space of square integrable functions on the measure space $(\underline{\underline{X}}, dx)$ and $C_{com}(\underline{\underline{X}})$, $C_0(\underline{\underline{X}})$ are the collections of real valued continuous functions on $\underline{\underline{X}}$ respectively with compact support and "vanishing at infinity." Questions of measurability are generally taken for granted-thus functions are usually understood to be measurable with respect to the obvious sigma algebra.

Table of Contents

Chapter I. General Theory

This chapter unifies and extends some of the results in [44] and [46].

In Section 1 we establish the connection between the submarkov property for symmetric semigroups P_t and the contractivity property for Dirichlet spaces $(\underline{\underline{F}}, E)$. This was first discovered by A. Beurling and J. Deny [1] but apparently it was M. Fukushima who first appreciated its significance for Markov processes. Also in Section 1 we introduce what seems to be the appropriate notion of "irreducibility" and we distinguish the transient and recurrent cases. We define the extended Dirichlet space $\underline{\underline{F}}_{(e)}$ by completing $\underline{\underline{F}}$ relative to the E form alone (that is, without adding a piece of the standard inner product) and we show that $(\underline{\underline{F}}_{(e)}, E)$ is an honest Hilbert space when $(\underline{\underline{F}}, E)$ is transient.

In Section 2 we show how a given Dirichlet space can be transformed into a regular one by introducing an appropriate modification of the state space. Our construction differs only slightly from Fukushima's in [21].

Some potential theory for regular Dirichlet spaces is developed in Section 3 and used in Section 4 to construct a "decent" Markov process. The main result was first established by Fukushima [22]. Our approach differs from his in that we avoid Ray resolvents and quasi-homeomorphisms.

In Section 5 we adapt G. A. Hunt's construction of "approximate Markov chains" to our situation.

In Section 6 we introduce various additive functionals, some of which are used to develop a theory of balayage in Section 7.

In Section 8 we study random time change. We show that the time changed process is symmetric relative to the "time changing measure" and we identify the time changed Dirichlet space. One immediate application is that if $(\underline{\underline{F}},E)$ is recurrent then the constant function 1 is in the extended space $\underline{\underline{F}}_{(e)}$ and the norm $E(1,1) = 0$. In particular $\underline{\underline{F}}_{(e)}$ is not a Hilbert space which complements the result in Section 1 for the transient case.

1. Transience and Recurrence

1.1. Definition. A symmetric (submarkovian) resolvent on $L^2(dx)$ is a family of bounded symmetric linear operators $G_u, u > 0$ on $L^2(dx)$ satisfying.

1.1.1. $G_u f \geq 0$ whenever $f \geq 0$ and $uG_u f \leq 1$ whenever $f \leq 1$.

1.1.2. $G_u - G_v = (v-u)\ G_u G_v$. ///

Of course the inequalities in 1.1.1 are in the almost everywhere sense. It follows from 1.1.1 together with the Cauchy-Schwarz inequality that each uG_u is a contraction on $L^2(dx)$. (Indeed this is the case on $L^p(dx), 1 \leq p \leq +\infty$.)

Consider now a given symmetric resolvent $\{G_u, u > 0\}$ and assume that the common range of the G_u is dense in $L^2(dx)$. It follows from 1.1.2 that there exists a unique non-positive definite self adjoint operator A on $L^2(dx)$ such that

$$G_u = (u - A)^{-1} \qquad u > 0. \tag{1.1}$$

For $t > 0$ let

$$P_t = e^{tA} \tag{1.2}$$

with the right side defined by the usual operator calculus. It is easy to check that the P_t satisfy the following conditions.

1.2.1. Each P_t is a symmetric contraction on $L^2(dx)$.

1.2.2. $P_t P_s = P_{t+s}$ for $s, t > 0$.

1.2.3. $\text{Lim}_{t \downarrow 0}\ P_t = 1$ in the strong operator topology on $L^2(dx)$.

1.2.4. $G_u = \int_0^\infty dt\ e^{-ut}\ P_t$ for $u > 0$.

1.2.5. $P_t f \geq 0$ whenever $f \geq 0$ and $P_t f \leq 1$ whenever $f \leq 1$. /// 1.2

Because of 1.2.3 there is no difficulty in interpreting the integral in 1.2.4 as a Riemann integral. To establish 1.2.5 it suffices to apply Laplace inversion as in [16, XIII. 4] to 1.1.1.

Let $\sqrt{-A}$ be the unique nonnegative definite self adjoint square root of $-A$ and put

(1.3) $$\underline{\underline{F}} = \text{domain } \sqrt{-A}$$

$$E(f, g) = \int dx \sqrt{-A}\ f(x) \sqrt{-A}\ g(x) \qquad f, g \text{ in } \underline{\underline{F}}.$$

Also for $u > 0$ put

(1.4) $$E_u(f, g) = E(f, f) + u \int dx\ f^2(x).$$

We will refer to the pair $(\underline{\underline{F}}, E)$ as the *associated Dirichlet space.* The following is quite useful for analyzing $(\underline{\underline{F}}, E)$.

Lemma 1.1. *For general f in $L^2(dx)$*

$$(1/t) \int dx \{f(x) - P_t f(x)\} f(x)$$

increases as $t \downarrow 0$ and

$$u \int dx \{f(x) - u G_u f(x)\} f(x)$$

increases as $u \uparrow \infty$. Each of these expressions remains bounded if and only if f is in $\underline{\underline{F}}$ and in this case

(1.5) $$E(f, f) = \operatorname{Lim}_{t \downarrow 0} (1/t) \int dx \{f(x) - P_t f(x)\} f(x)$$

(1.6) $$E(f, f) = \operatorname{Lim}_{u \uparrow \infty} u \int dx \{f(x) - u G_u f(x)\} f(x). ///$$

Proof. By the spectral theorem there exists a measure space 1.3
$(\underline{X}^{\sim}, dx^{\sim})$, an isometry $\mathcal{U}$. $L^2(\underline{X}, dx) \to L^2(\underline{X}^{\sim}, dx^{\sim})$, and a nonpositive measurable function $A^{\sim}$ on $\underline{X}^{\sim}$ such that f in $L^2(\underline{X}, dx)$ belongs to $\underline{F}$ if and only if

$$(1.7) \qquad - \int dx^{\sim} A^{\sim}(x) \{ \mathcal{U} f(x^{\sim}) \}^2$$

is finite in which case $E(f, f)$ is given by (1.7). Also

$$(1.8) \qquad \int dx \{f(x) - P_t f(x)\} f(x) = \int dx^{\sim} \{1 - e^{tA^{\sim}(x^{\sim})}\} \{ \mathcal{U} f(x^{\sim})\}^2$$

$$\int dx \{f(x) - u G_u f(x)\} f(x) = \int dx^{\sim} \{ -A^{\sim}(x^{\sim})/(u - A^{\sim}(x^{\sim})) \} \{\mathcal{U} f(x^{\sim})\}^2 .$$

The lemma follows since for $a > 0$ the expressions $ua/(u+a)$ and $(1/t)(1-e^{-ta})$ increase to a as $u \uparrow \infty$ and $t \downarrow 0$. For the first expression it suffices to observe that if $\varphi(u) = ua/(u+a)$ then $\varphi'(u) = \{a(u+a) - ua\} / (u+a)^2 = a^2/(u+a)^2$. For the second expression let $\varphi(t) = (1/t)(1-e^{-ta})$. Then $\varphi'(t) = \{tae^{-ta} - 1 + e^{-ta}\} / t^2$ and it suffices to show that $\psi(b) = be^{-b} + e^{-b} - 1$ is negative for $b > 0$. Clearly $\psi(+\infty) = -1$ and for $b > 0$ sufficently small $\psi(b) \sim b(1-b) + (1 - b + \frac{1}{2}b^2) - 1 = -\frac{1}{2} b^2$ is negative. Finally $\psi'(b) = e^{-b} - be^{-b} - e^{-b} = -be^{-b}$ is negative and we are done. ///

It is elementary that $\underline{F}$ is a Hilbert space relative to any of the inner products E_u. A deeper result can be established with the help of Lemma 1.1. We say that g is a normalized contraction of f if there exist everywhere defined versions of f and g such that

$$(1.9) \qquad |g(x)| \le |f(x)| \; ; \quad |g(x) - g(y)| \le |f(x) - f(y)| .$$

With our hypotheses on $\underline{X}$ there exists for $u > 0$ a symmetric measure

$G_u(dx,dy)$ on $\underline{\underline{X}} \times \underline{\underline{X}}$ such that

$$\int dx\, f(x)\, G_u g(x) = \iint G_u(dx,dy)\, f(x) g(y)$$

for f, g in $L^2(dx)$. Then for f in $\underline{\underline{F}}$ and for g a normalized contraction of f

$$(1.10) \qquad \int dx \{g(x) - uG_u g(x)\}\, g(x)$$

$$= \int dx\, g^2(x)\, \{1 - uG_u 1(x)\} + u \iint G_u(dx,dy)\, \{g^2(x) - g(x)g(y)\}$$

$$= \int dx\, g^2(x)\, \{1 - uG_u 1(x)\} + \tfrac{1}{2} u \iint G_u(dx,dy)\, \{g(x) - g(y)\}^2$$

and it follows that

$$\int dx\, \{g - G_u g(x)\}\, g(x) \leq \int dx \{f - G_u f(x)\}\, f(x).$$

Thus by Lemma 1.1 g belongs to $\underline{\underline{F}}$ and $E(g,g) \leq E(f,f)$. We summarize in

<u>1.3.1.</u> $\underline{\underline{F}}$ is a Hilbert space relative to any of the inner products (1.4).

<u>1.3.2.</u> If f is in $\underline{\underline{F}}$ and if g is a normalized contraction of f then also g belongs to $\underline{\underline{F}}$ and $E(g,g) \leq E(f,f)$. ///

In general any pair $(\underline{\underline{F}}, E)$ with $\underline{\underline{F}}$ a dense linear subset of $L^2(dx)$ and with E a bilinear form on F will be called a <u>Dirichlet space on</u> $L^2(dx)$ if it satisfies 1.3.1 and 1.3.2. We have shown above that the Dirichlet space associated with a symmetric submarkovian resolvent on $L^2(dx)$ is a Dirichlet space on $L^2(dx)$. We now prove conversely that every Dirichlet space $(\underline{\underline{F}}, E)$ on $L^2(dx)$ is associated with a unique submarkovian resolvent.

Let $(\underline{\underline{F}}, E)$ be a Dirichlet space on $L^2(dx)$. For $u > 0$ let G_u be the unique bounded operator from $L^2(dx)$ into $\underline{\underline{F}}$ determined by

(1.11) $$E_u(G_u\varphi, g) = \int dx\, \varphi(x)\, g(x)$$

for φ in $L^2(dx)$ and g in $\underline{\underline{F}}$. It is easy to verify the resolvent identity 1.1.2 and that for $u > 0$ the operator uG_u is a symmetric contraction on $L^2(dx)$. In place of 1.1.1 we prove the following more general result. Let T be any mapping from the reals to the reals satisfying

(1.12) $$T0 = 0 \;; \quad |T\alpha - T\beta| \le |\alpha - \beta| .$$

We will show that if $T\varphi = \varphi$ then also $Tu\,G_u\varphi = u\,G_u\varphi$. (Then 1.1.1 follows from the special cases $T\alpha = \alpha^+$ and $T\alpha = \min(\alpha, 1)$.) Fix one such φ and note that for f in $\underline{\underline{F}}$ and $u > 0$

$$\begin{aligned} &E_u(f - u\,G_u\varphi,\ f - u\,G_u\varphi) \\ &= E(f,f) + u\int dx\,\{ f^2(x) + u\,\varphi(x)\,G_u\varphi(x) - 2f(x)\varphi(x)\} \\ &= E(f,f) + u\int dx\,\{f(x) - \varphi(x)\}^2 + u\int dx\,\{u\,\varphi(x)\,G_u\varphi(x) - \varphi^2(x)\}. \end{aligned}$$

and therefore the functional

$$\Phi(f) = E(f,f) + u\int dx\,\{f(x) - \varphi(x)\}^2$$

has the unique minimum $f = u\,G_u\varphi$. But in general $\Phi(Tf) \le \Phi(f)$ and so $T\,uG_u\varphi = u\,G_u\varphi$. We summarize these results in

<u>Theorem 1.2.</u> <u>Let</u> $\{G_u, u > 0\}$ <u>be a submarkovian resolvent on</u> $L^2(dx)$ <u>with common range dense in</u> $L^2(dx)$. <u>Then the pair</u> $(\underline{\underline{F}}, E)$ <u>defined by (1.3) is a Dirichlet space on</u> $L^2(dx)$. <u>Conversely every Dirichlet space on</u> $L^2(dx)$ <u>is associated in this way with a unique submarkovian resolvent on</u> $L^2(dx)$ <u>with dense common range.</u> ///

<u>Remark 1.</u> Our restriction to resolvents having dense range and therefore to Dirichlet spaces $\underline{\underline{F}}$ which are dense in $L^2(dx)$ is not necessary for Theorem 1.2. In the general case we need only replace $L^2(dx)$ by the

closure of the range of the resolvent operators. Also our assumptions about $\underline{X}$ can be weakened considerably. It suffices for example that $\underline{X}$ be an absolute Borel set in the sense of [39]. ///

Remark 2. The connection between a submarkovian resolvent G_u, $u > 0$ and a submarkovian semigroup P_t, $t > 0$ is a familiar tool in the theory of Markov processes. To our knowledge it was first used systematically by W. Feller in his fundamental papers on diffusions and solutions of the Kolmogorov equations. The connection between symmetric resolvents and Dirichlet spaces is less well known. The basic idea goes back to Beurling and Deny [1]. However it seems that M. Fukushima [20] first appreciated its significance in the context of Markov processes. ///

We continue to work now with a Dirichlet space $(\underline{F}, E)$, as in Theorem 1.2 and its associated resolvent $\{G_u, u > 0\}$ and semigroup $\{P_t, t > 0\}$. For general $f \geq 0$ define

$$P_t f = \text{Lim}\, P_t f_n$$

with f_n any sequence of square integrable functions which increase to f almost everywhere. Also define

$$P_t f = P_t f^+ - P_t f^-$$

when the right side converges. It then follows from symmetry and from 1.2.5 that the extended operators are contractions on $L^p(dx)$, $1 \leq p \leq +\infty$. Extend the operators uG_u in a similar way and note that by 1.1.1 the uG_u are contractions on $L^p(dx)$, $1 \leq p \leq +\infty$.

Lemma 1.3. For f in $L^1(dx)$ 1.7

$$\operatorname{Lim}_{t \downarrow 0} \int dx \, | f(x) - P_t f(x) | = 0. \;///$$

Proof. We use Rota's well known device [41]. Let Ω be the set of maps ω from the half line $[0, \infty)$ into the augmented space $\underline{X} \cup \{\partial\}$, let X_t, $t \geq 0$ be the usual trajectory variables and let $\mathcal{P}$ be the unique measure on Ω such that

$$\int \mathcal{P}(d\omega)\, f_0(X_0(\omega)) \cdots f_n(X_{t_n}(\omega))$$
$$= \int dx\, f_0(x)\, P_{t_1} f_1 \cdots P_{t_n - t_{n-1}} f_n(x)$$

for $0 < t_1 < \cdots < t_n$ and for $f_0, \ldots, f_n \geq 0$ on $\underline{X}$. Let $\mathcal{T}_0$ be the σ-algebra on Ω generated by X_0 and for $t > 0$ let $\mathcal{T}_t$ be the σ-algebra generated by X_s, $s \geq t$. For f in $L^1(X)$

$$\mathcal{T}_0 \mathcal{T}_t\, f(X_0) = P_t(1/P_t 1) P_t f(X_0) \geq P_{2t} f(X_0)$$

and it follows that the functions $P_{2t} f(X_0)$, $t > 0$ are uniformly integrable on Ω in the sense of Hunt [27, Section I.6] and therefore the functions $P_t f(x), t > 0$ are uniformly integrable on $\underline{X}$. But this proves the lemma because it follows from 1.2.3 that $P_t f \to f$ in measure as $t \downarrow 0$ for f in $L^1(dx) \cap L^2(dx)$. ///

Remark. The same argument gives an L^p result for $1 \leq p < +\infty$. Also it follows from symmetry that P_t is continuous in t with respect to the weak* topology on $L^\infty(dx)$. Once this is established it is routine to extend 1.2.4 to $L^p(dx)$, $1 \leq p \leq +\infty$. ///

A subset A of $\underline{X}$ is proper invariant if neither A nor its complement $A^c = \underline{X} - A$ is dx null and if $P_t 1_A \leq 1_A$ for all $t > 0$. It follows from symmetry that A is proper invariant if and only if its complement A^c is. Moreover if f belongs to $L^2(dx)$ then

$$\int dx\, f(x)\{f(x) - P_t f(x)\} = \int dx\, 1_A f(x)\{1_A f(x) - P_t 1_A f(x)\}$$

$$+ \int dx\, 1_{A^c} f(x)\{1_{A^c} f(x) - P_t 1_{A^c} f(x)\}$$

$$\geq \int dx\, 1_A f(x)\{1_A f(x) - P_t 1_A f(x)\}.$$

Thus if f belongs to $\underline{\underline{F}}$ then so does $1_A f$ and we can obtain a new Dirichlet space by restricting everything to A. This suggests that an appropriate generalization of irreducibility for Markov chains is

1.4. Condition of Irreducibility. There exist no proper invariant sets.///

From now on we assume that this condition is satisfied. Our feeling is that this restriction will be harmless in practice.

Next we apply the techniques associated with the Hopf decomposition to distinguish the transient and recurrent cases. Our source for these techniques is the book of Foguel [17]. We begin by adapting Garsia's well known argument in the discrete time case. Let f be in $L^1(dx)$ and for $n \geq 0$ define

$$(1.13) \qquad S_n f(x) = \int_0^{2^{-n}} dt\; P_t f(x)$$

$$S^*_{n,m} f(x) = \max_{1 \leq k \leq m} \int_0^{k2^{-n}} dt\; P_t f(x)$$

$$S^*_n f(x) = \sup_{1 \leq k} \int_0^{k2^{-n}} dt\; P_t f(x).$$

The integrals are well defined in the L^1 sense because of Lemma 1.3. Clearly

$$\int_0^{k2^{-n}} dt\, P_t f = S_n f + P_{2^{-n}} \int_0^{(k-1)2^{-n}} dt\, P_t f$$

and therefore

$$S^*_{n,m} f \leq S_n f + P_{2^{-n}} (S^*_{n,m} f)^+$$

Thus

$$(1.14) \qquad \int_{[S^*_{n,m} f > 0]} dx\, S_n f(x)$$

$$\geq \int dx\, (S^*_{n,m} f)^+ - \int dx\, P_{2^{-n}} (S^*_{n,m} f)^+ \geq 0$$

and after passage to the limit $m \uparrow \infty$

$$(1.15) \qquad \int_{[S^*_n f > 0]} dx\, S_n f(x) \geq 0.$$

Since $2^n S_n f \to f$ in L^1 as $n \uparrow \infty$ also

$$2^n 1_{[S^*_n f > 0]} S_n f \to 1_E f$$

in L^1 where $E = \bigcup_{n=0}^{\infty} [S^*_n f > 0]$. (Again we are using the concept of uniform integrability as formulated in [27, I.6].) Thus after multiplying (1.15) by 2^n and passing to the limit $n \uparrow \infty$ we obtain

<u>Lemma 1.4.</u> <u>Let f be in $L^1(dx)$ and let</u>

$$E = \{x : \sup_{n \geq 0,\, k \geq 1} \int_0^{k2^{-n}} dt\, P_t f(x) > 0\}.$$

Then $\int_E dx\, f(x) \geq 0$. ///

We define the Green's operator G by

(1.16) $$Gf = \int_0^\infty dt\; P_t f$$

when it makes sense.

Corollary 1.5. Let f, g > 0 almost everywhere be in $L^1(dx)$. Then

$$[Gf < +\infty] = [Gg < +\infty] \quad \text{almost everywhere.} \;///$$

Proof. Let $A = [Gf = \infty,\; Gg < +\infty]$. Then for a > 0

$$B = \{\, x : \sup_{n \geq 0,\; k \geq 1} \int_0^{k2^{-n}} dt\; P_t(f-ag)\,(x) > 0 \,\}$$

contains A and therefore by Lemma 1.4

$$\int_B dx\, \{ f(x) - ag(x) \} \geq 0$$

and 'a fortiori'

$$a \int_A dx\, g(x) \leq \int dx\, f(x).$$

The corollary follows upon letting $a \uparrow \infty$. ///

For $f \geq 0$ and nontrivial in $L^1(dx)$ the set $[Gf = 0]$ is clearly invariant and therefore must be null by our condition of irreducibility. Similarly $[Gf < +\infty]$ is invariant and therefore either this set or its complement $[Gf = +\infty]$ is null. By Corollary 1.5 the outcome is independent of the choice of $f > 0$ almost everywhere. But if $f \geq 0$ is nontrivial then by irreducibility $G_1 f > 0$ almost everywhere and since $Gf \geq G G_1 f$ it follows that the outcome is independent of the choice of $f \geq 0$ and nontrivial in $L^1(dx)$. Thus the following makes sense.

1.5. Definition. The Dirichlet space $(\underline{F}, E)$ is transient if

$$Gf \text{ finite almost everywhere}$$

for all f in $L^1(dx)$ and recurrent if

$$Gf = +\infty \qquad \text{almost everywhere}$$

for all $f \geq 0$ and nontrivial in $L^1(dx)$. ///

For $h > 0$

$$\int_0^t ds\, P_s(1-P_h 1) = \int_0^h ds\, P_s 1 - \int_t^{t+h} ds\, P_s 1$$

stays bounded as $t \uparrow \infty$. There follows

Theorem 1.6. If $(\underline{F}, E)$ is recurrent, then $P_t 1 = 1$ almost everywhere for every $t > 0$. ///

Of course the converse to Theorem 1.6 is false. (Consider for example standard Brownian motion in R^d for $d \geq 3$.)

1.6 Definition. f belongs to the extended Dirichlet space $\underline{F}_{(e)}$ if there exists a sequence f_n is $\underline{F}$ such that

1.6.1. $\{f_n\}$ is Cauchy relative to E.

1.6.2. $f_n \to f$ almost everywhere on $\underline{X}$. ///

Condition 1.6.1 can be replaced by the apparently weaker

1.6.1' $E(f_n, f_n)$ is bounded independent of n. ///

To prove this we adapt the proof of a well known theorem of Banach and Saks [40, p.80]. We assume that $\{f_n\}$ satisfies 1.6.1' and we show that the Cesàro means of a subsequence satisfy 1.6.1. To avoid excessive computation we temporarilly introduce $\mathcal{J}$ the Hilbert space formed from $\underline{F}$ by first identifying functions in $\underline{F}$ whose difference has E norm zero and then completing. We use the same symbol for functions in $\underline{F}$ and their corresponding equivalence classes in $\mathcal{J}$. After selecting a subsequence we can assume that $f_n \to \varphi$ weakly in $\mathcal{J}$ and after again selecting a subsequence we can assume that

$$E(\varphi - f_m, \varphi - f_n) < (1/n) \quad \text{for} \quad n > m.$$

Clearly 1.6.1 will follow if we show that

$$(1.17) \qquad E(\varphi - \{f_1 + \cdots + f_n\}/n, \varphi - \{f_1 + \cdots + f_n\}/n) \to 0.$$

But the left side of (1.17)

$$= (1/n^2) \ \Sigma_{k=1}^{n} E(\varphi - f_k, \varphi - f_k)$$

$$+ 2(1/n^2) \ \Sigma_{k=2}^{n} \ \Sigma_{\ell=1}^{k-1} E(\varphi - f_k, \varphi - f_\ell)$$

$$\leq 4 (1/n) \ \sup_m E(f_m, f_m) + 2(1/n^2) \ \Sigma_{k=2}^{n} (1/k)(k-1)$$

$$\leq (4/n) \ \sup_m E(f_m, f_m) + (2/n)$$

and we are done.

<u>Remark.</u> There is an alternative noncomputational argument which uses the fact that the weak closure of a convex subset of a Hilbert space is also the strong closure. ///

Lemma 1.7. *Let f be in $\underline{\underline{F}}_{(e)}$ and let $\{f_n\}$ be as in Definition 1.6.*

(i) *The limit* $\mathrm{Lim}_{n\uparrow\infty} E(f_n, f_n)$ *is independent of the choice of the approximating sequence* $\{f_n\}$. *Therefore E extends uniquely to* $\underline{\underline{F}}_{(e)}$ *by continuity and*

$$(1.18)\qquad E(f,f) = \mathrm{Lim}_{n\uparrow\infty} E(f_n, f_n).$$

(ii) *The expressions*

$$(1.19)\qquad (1/t)\int dx\, \{1-P_t 1(x)\}\, f^2(x) + \tfrac{1}{2}(1/t)\iint P_t(dx,dy)\{f(y)-f(x)\}^2$$

$$(1.19')\qquad u\int dx\, \{1-uG_u 1(x)\}\, f^2(x) + \tfrac{1}{2}\, u^2\iint G_u(dx,dy)\{f(y)-f(x)\}^2$$

are finite for $u,t>0$ *and increase to* $E(f,f)$ *as* $t\downarrow 0$ *and* $u\uparrow\infty$ *respectively.*///

Proof. By the triangle inequality $\ell = \mathrm{Lim}_{n\uparrow\infty} E(f_n, f_n)$ exists. Thus (i) will follow from (ii) if we show that the expressions (1.19) and (1.19') increase to ℓ. We give the argument only for (1.19) and temporarilly introduce the special notation $E_t(f,f)$ for (1.19). The estimate

$$E_t(f_m-f_n, f_m-f_n) \le E(f_m-f_n, f_m-f_n)$$

is valid by Lemma 1.1 and it follows with the help of Fatou's lemma that

$$E_t(f-f_n, f-f_n) \le \mathrm{Lim\,sup}_{m\uparrow\infty} E(f_m-f_n, f_m-f_n).$$

By the triangle inequality $E_t(f,f)$ is finite and

$$|E_t^{\frac{1}{2}}(f,f) - E_t^{\frac{1}{2}}(f_n, f_n)| \le \mathrm{Lim\,sup}_{m\uparrow\infty} E^{\frac{1}{2}}(f_m-f_n, f_m-f_n).$$

In particular $E_t(f_n, f_n) \to E_t(f, f)$ and so again by Lemma 1.1 the expressions $E_t(f, f)$ increase as $t \downarrow 0$ and are dominated by ℓ. Finally convergence to ℓ follows from the estimate

$$\ell - E_t(f, f)$$

$$\leq | \ell - E(f_n, f_n)| + | E(f_n, f_n) - E_t(f_n, f_n)| + | E_t(f_n, f_n) - E_t(f, f)|$$

$$\leq 2 \text{ Lim sup}_{m \uparrow \infty} E(f_m - f_n, f_m - f_n) + | E(f_n, f_n) - E_t(f_n, f_n) | . ///$$

Remark 1. One important corollary of Lemma 1.7-(ii) is that $\underline{\underline{F}} = \underline{\underline{F}}_{(e)} \cap L^2(dx)$.///

Remark. 2. If the expressions (1.19) or (1.19') are uniformly bounded for f and if f is dominated by g in $\underline{\underline{F}}_{(e)}$, then f is in $\underline{\underline{F}}_{(e)}$. This can be proved by restricting attention to $f \geq 0$ and considering the approximations $\min(f, g_n)$ for f with g_n chosen for g as in Definition 1.6. Clearly estimates for (1.19) or (1.19') alone cannot be sufficient. If $(\underline{\underline{F}}, E)$ is transient and conservative then (1.19) and (1.19') vanish for $f = 1$ but 1 is not in $\underline{\underline{F}}_{(e)}$. With the results in Chapter II it can be shown that (1.19) or (1.19') alone do characterize the reflected space $\underline{\underline{F}}_{ref}$ defined in Section 14. ///

Lemma 1.8. Let $(\underline{\underline{F}}, E)$ be transient. Then $\text{Lim}_{t \uparrow \infty} P_t = 0$ in the strong operator topology on $L^2(dx)$.///

Proof. It follows from the spectral theorem that if the lemma is false then there exists f in $L^2(dx)$ such that $P_t f = f$ for all $t > 0$. Therefore it suffices to observe that for f in $L^1(dx) \cap L^2(dx)$ and for $s > 0$

$$\text{Lim}_{t \uparrow \infty} \int_t^{t+s} du\, P_u f = 0$$

almost everywhere. ///

We suppose now that $(\underline{F}, E)$ is transient and temporarily introduce the notation

$$S_t = \int_0^t du\, P_u .$$

Fix $\varphi \geq 0$ in $L^1(dx) \cap L^2(dx)$ such that $G\varphi \leq M < +\infty$ almost everywhere on the set where $\varphi > 0$. For $s, t > 0$

$$(1/s)\,(1-P_s)\, S_t\, \varphi = (1/s) \int_0^s du\, P_u \varphi - (1/s) \int_t^{t+s} du\, P_u \varphi$$

and it follows from Lemma 1.1 that $S_t \varphi$ belongs to $\underline{F}$ and that

(1.20) $$E(S_t \varphi, f) = \int dx\, f(x)\, \{\varphi(x) - P_t\, \varphi(x)\}$$

for f in $\underline{\underline{F}}$ and so by Lemma 1.8

(1.21) $$\int dx\, f(x)\, \varphi(x) = \text{Lim}_{t \uparrow \infty} E(S_t\, \varphi, f).$$

A special case of (1.20) is

(1.22) $$\begin{aligned} E(S_t\, \varphi, S_t \varphi) &= \int dx\, S_t\, \varphi(x)\, \{\varphi(x) - P_t \varphi(x)\} \\ &\leq \int dx\, G\varphi(x)\, \varphi(x) \\ &\leq M\, \|\varphi\|_1 . \end{aligned}$$

Thus $G\varphi$ is in $\underline{\underline{F}}_{(e)}$ and then by (1.21)

(1.23) $$E(G\varphi, f) = \int dx\, f(x)\, \varphi(x)$$

for f in $\underline{\underline{F}}$. This extends immediately to f in $\underline{\underline{F}}_{(e)}$ and to $\varphi \geq 0$ in $L^1(dx)$ such that $G\varphi$ is bounded on the set where $\varphi > 0$. Also it follows upon varying φ in (1.23) that if $\{f_n\}_{n=1}^{\infty}$ is Cauchy in $\underline{\underline{F}}_{(e)}$ then a subsequence converges almost everywhere. Of course the limit function f is in $\underline{\underline{F}}_{(e)}$ and $f_n \to f$ relative to 'E. We summarize in

Theorem 1.9. If $(\underline{\underline{F}}, E)$ is transient then $\underline{\underline{F}}_{(e)}$ is a Hilbert space with E as inner product. Moreover if $\varphi \geq 0$ in $L^1(dx)$ is such that $G\varphi$ is bounded on the set where $\varphi > 0$ then $G\varphi$ is in $\underline{\underline{F}}_{(e)}$ and (1.23) is valid for f in $\underline{\underline{F}}_{(e)}$. ///

Remark 1. We will see in Section 3 that if $G\varphi \leq M$ almost everywhere on the set where $\varphi > 0$ then actually $G\varphi \leq M$ almost everywhere.///

Remark 2. If $(\underline{\underline{F}}, E)$ is recurrent and if dx is bounded then by Theorem 1.6 the function 1 is in $\underline{\underline{F}}$ and $E(1,1) = 0$. Thus $\underline{\underline{F}}_{(e)}$ cannot be made into a Hilbert space relative to E. This is also true for dx unbounded, but the proof must wait until Section 8. ///

Remark 3. It follows from the spectral theorem that the operators P_t and $u\, G_u$ are contractive on $\underline{\underline{F}}$ relative to the E norm. Therefore in the transient case these operators extend by continuity to contractions on $\underline{\underline{F}}_{(e)}$. From now on we take these extensions for granted. Also P_t, $u\, G_u \to 1$ in the strong operator topology on $\underline{\underline{F}}_{(e)}$ as $t \downarrow 0$ and $u \uparrow \infty$, the identity $G_u = \int_0^{\infty} dt\, e^{-ut}\, P_t$ is valid for the extension, and $P_t \to 0$ in the strong operator topology on $\underline{\underline{F}}_{(e)}$ as $t \uparrow \infty$. ///

Remark 4. We have recently received a preprint from M. Fukushima entitled "Almost polar sets and an ergodic theorem." Among other things this paper deals with transience and recurrence in a nonsymmetric setting. ///

2. Regular Dirichlet Spaces

2.1. Definition. The Dirichlet space $(\underline{F}, E)$ on $L^2(dx)$ is regular if

2.1.1. $\underline{F} \cap C_{com}(\underline{X})$ is uniformly dense in $C_{com}(\underline{X})$ and E_1 dense in $\underline{F}$.

2.1.2. dx is everywhere dense. That is $\int_G dx > 0$ for any nonempty open set G. ///

Condition 2.1.2 is harmless since we can always replace $\underline{X}$ by the support of dx. Notice that if $(\underline{F}, E)$ is regular and transient then also $\underline{F} \cap C_{com}(\underline{X})$ is E dense in $\underline{F}_{(e)}$.

We consider now a Dirichlet space $(\underline{F}, E)$ on $L^2(dx)$ such that $\underline{F}$ is dense in $L^2(dx)$ and we construct a regular Dirichlet space by modifying the state space $\underline{X}$. Our construction differs only slightly from one first given by Fukushima [21].

This section is essentially a repetition of the appendix in [44].

Note first that if f in $\underline{F}$ is bounded almost everywhere, then f^2 belongs to $\underline{F}$ and so the subcollection of f in $\underline{F}$ which are integrable and bounded form an algebra. Obviously this algebra is dense in $\underline{F}$ and since $\underline{F}$ itself is separable (since it is the domain of a self adjoint operator on a separable Hilbert space) there exists a subset $\underline{B}_0$ of $\underline{F}$ satisfying

2.2.1. $\underline{B}_0$ is countable

2.2.2. $\underline{B}_0$ is an algebra over the rationals

2.2.3. Every f in $\underline{B}_0$ is integrable and bounded.

2.2.4. $\underline{B}_0$ is dense in $\underline{F}$ (and therefore in $L^2(dx)$). ///

The uniform closure $\underline{B}$ is a commutative Banach algebra and so the well known techniques associated with the Gelfand transform of $\underline{B}$ are available.

2.2

Let $\underline{\underline{Y}}$ be the collection of real valued functions γ on $\underline{\underline{B}}$, not identically zero, which satisfy for f, g in $\underline{\underline{B}}$ and for a, b rational

2.3.1. $\gamma(f) \leq \|f\|_{\infty}$

2.3.2. $\gamma(fg) = \gamma(f)\,\gamma(g)$

2.3.3. $\gamma(af + bg) = a\,\gamma(f) + b\,\gamma(g)$. ///

We give $\underline{\underline{Y}}$ the weakest topology which makes continuous all of the real valued functions F on $\underline{\underline{Y}}$ which can be represented $F(\gamma) = \gamma(f)$ for some f in $\underline{\underline{B}}$. It is well known and easy to verify directly that $\underline{\underline{Y}}$ is then a separable locally compact Hausdorff space which is compact if and only if 1 is in $\underline{\underline{B}}$.

Once and for all choose everywhere defined versions for f in $\underline{\underline{B}}_0$ and let $\underline{\underline{X}}_0$ be the subset of x in $\underline{\underline{X}}$ satisfying

2.4.1. $|f(x)| \leq \|f\|_{\infty}$ for f in $\underline{\underline{B}}_0$.

2.4.2. $(fg)(x) = f(x)g(x)$ and $(f+g)(x) = f(x) + g(x)$ for f, g in $\underline{\underline{B}}_0$.

2.4.3. $(af)(x) = af(x)$ for a rational and for f in $\underline{\underline{B}}_0$. ///

Clearly $\underline{\underline{X}}_0$ is a Borel subset of $\underline{\underline{X}}$ with full measure. There exists a unique mapping $j : \underline{\underline{X}}_0 \to \underline{\underline{Y}}$ such that

$$(jx)(f) = f(x)$$

for f in $\underline{\underline{B}}_0$ and for x in $\underline{\underline{X}}_0$. Clearly j is Borel measurable and so there exists a unique Borel measure $d\gamma$ on $\underline{\underline{Y}}$ such that

$$\int_{Y} d\gamma\, \varphi(\gamma) = \int_{\underline{\underline{X}}_0} dx\, \varphi(jx)$$

for nonnegative φ on $\underline{\underline{Y}}$. It follows from the integrability of φ in B_0 that $d\gamma$ is Radon. We use the same symbol j to denote the natural mapping of $\underline{\underline{B}}$ onto $C_0(\underline{\underline{Y}})$. Thus

$$jf(\gamma) = \gamma(f) \qquad f \in \underline{\underline{B}},\ \gamma \in \underline{\underline{Y}}.$$

Clearly for f in $\underline{\underline{B}}$ and for any polynomial P

$$jP(f) = P(jf)$$

and since any T satisfying (1. 12) is continuous and therefore can be approximated uniformly by polynomials on compact sets, also

$$jT(f) = T(jf).$$

For $f, g \geq 0$ in $\underline{\underline{B}}$

$$\begin{aligned}
&\int d\gamma\ jf(\gamma)\ jg(\gamma)\\
&\quad = \int dx\ jf(jx)\ jf(jx)\\
&\quad = \int dx\ jx(f)\ j\,x(g)\\
&\quad = \int dx\ jx(fg)\\
&\quad = \int dx\ f(x)g(x)
\end{aligned}$$

and it follows that j is an isometry from $L^2(dx) \cap \underline{\underline{B}}$ onto $L^2(\gamma) \cap C_0(\underline{\underline{Y}})$. Let jE be defined on $j\underline{\underline{B}}_0$ by

$$jE(jf, jg) = E(f, g).$$

The desired Dirichlet space is the closure of jB_0 relative to the inner product

$$jE_1(f, f) = jE(f, f) + \int d\gamma\ f^2(\gamma)$$

together with the continuous extension of E to this closure. To establish regularity for this Dirichlet space it only remains to check that $d\gamma$ is dense. For this it suffices to show that

$$(2.1) \qquad \text{meas. } \bigcap_{i=1}^{n} \{x : | f_i(x) - \gamma(f_i)) < \varepsilon \} > 0$$

for any choice of $\varepsilon > 0$ of γ in $\underline{\underline{Y}}$ and of $f_1, \ldots, f_n$ in $\underline{\underline{B}}$. If (2.1) is false then there exist polynomials P_m in n indeterminates such that $P_m(f_1, \ldots, f_n)$ converges uniformly to

$$g = \min_{i=1}^{n} | f_i - \gamma(f_i) |^{-1}$$

and it follows that gh belongs to $\underline{\underline{B}}$ whenever h does. Since every h in $\underline{\underline{B}}$ can be represented

$$h = hg \max_{i=1}^{n} | f_i - \gamma(f_i) |$$

and since $\max_{i=1}^{n} | f_i - \gamma(f_i) |$ can be uniformly approximated by polynomials in $\{f_i - \gamma(f_i)\}_{i=1,\ldots,n}$ not containing the constant term, it follows that $\gamma(h) = 0$ for all h in $\underline{\underline{B}}$, which possibility has been ruled out by hypothesis.

Remark 1. The proof of (2.1) given on page 71 of [44] is incorrect. ///

Remark 2. Clearly the final state space $\underline{\underline{Y}}$ depends on the choice of $\underline{\underline{B}}_0$. However by Theorem 2.1 in [22] any $\underline{\underline{Y}}'$ resulting from a different choice for $\underline{\underline{B}}_0$ is related to $\underline{\underline{Y}}$ be a "capacity preserving quasi-homeomorphism" (see [22] for the precise definition) which is enough to guarantee that $\underline{\underline{Y}}$ and $\underline{\underline{Y}}'$ are identical from the point of view of the processes constructed below. We return to this subject in Section 19. ///

3. Some Potential Theory

Throughout this section $(\underline{F}, E)$ is a transient regular Dirichlet space on $L^2(dx)$. (The recurrent case can always be handled by replacing E with E_1.) The point of view taken here goes back at least to H. Cartan [3,4] for the classical Dirichlet spaces associated with the Laplacian and Brownian motion. The general formulation in terms of regular Dirichlet spaces is due to A. Beurling and J. Deny. (See [1].) The results themselves were first established by Fukushima [22] using an indirect approach.

This section differs only slightly from Section 1 in [44].

3.1. Definition. f in $\underline{F}_{(e)}$ is a potential if $E(f, g) \geq 0$ whenever g is in $\underline{F}_{(e)}$ and $g \geq 0$ almost everywhere. ///

Lemma 3.1. The following are equivalent for f in $\underline{F}_{(e)}$.

(i) f is a potential.

(ii) There exists a Radon measure μ such that

$$E(f, g) = \int \mu(dx)\, g(x)$$

for g in $\underline{F}_{(e)} \cap C_{com}(\underline{X})$.

(iii) $E(f+g, f+g) > E(f, f)$ whenever g is in $\underline{F}_{(e)}$ and $g \geq 0$.

(iv) $f \geq u\, G_u f$ for all $u > 0$.

(v) $f \geq P_t f$ for all $t > 0$. ///

Proof. That (ii) implies (i) is trivial. To prove that (i) implies (ii) let f be a potential and consider the nonnegative linear functional I defined on g in $\underline{F} \cap C_{com}(\underline{X})$ by $I(g) = E(f, g)$. If g_n decrease to 0 pointwise then by Dini's theorem they do so uniformly and after comparing to a fixed nonnegative g in $\underline{\underline{F}} \cap C_{com}(\underline{\underline{X}})$ which is ≥ 1 on the support of g_1 we see that $I(g_n) \downarrow 0$. Thus (ii) follows by the Daniell approach to integration (as presented for example in [34]). That (i) implies (iii) follows from

$$E(f + tg, f + tg) = E(f, f) + 2t\, E(f, g) + t^2\, E(g, g) \tag{3.1}$$

for $t = 1$ and that (iii) implies (i) follows from (3.1) for $t > 0$ sufficiently small. Equivalence of (iv) and (v) is easily established with the help of Laplace inversion. That (i) implies (iv) follows from

$$\begin{aligned} \int dx\, \{ f(x) - u\, G_u f(x) \}\, \varphi(x) \\ &= E(f - u\, G_u f,\, G\varphi) \\ &= E(f,\, G_u \varphi) \end{aligned} \tag{3.2}$$

which is valid for $\varphi \geq 0$ as in Theorem 1.9. That (v) implies (i) follows from

$$E\left(\int_0^s du\, P_u f, g\right) = \int ds\, \{ f(x) - P_s f(x) \}\, g(x) \tag{3.3}$$

upon dividing by s and passing to the limit $s \downarrow 0$. (The identity (3.3) follows in the same way as (1.20).) ///

Corollary 3.2. (i) Every potential is nonnegative.

(ii) If f, g are potentials, then so is $\min(f, g)$.

(iii) If f is a potential then so is $\min(f, c)$ for $c \geq 0$. ///

Proof. (i) follows since

$$\begin{aligned} E(|f|, |f|) &= E(f, f) + E(|f| - f, |f| - f) + 2E(f, |f| - f) \\ &\geq E(f, f) + E(|f| - f, |f| - f) \\ &\geq E(|f|, |f| + E(|f| - f, |f| - f) \end{aligned}$$

and therefore $E(|f| - f, |f| - f) = 0$. Conclusions (ii) and (iii) follow directly from Lemma 3.1- (iv) or (v). ///

Lemma 3.3. Let g be a potential in F and let f satisfy $0 \leq f \leq g$ and in addition

$$\text{either} \quad P_t f \leq f \qquad \text{all } t > 0$$

$$\text{or} \quad u G_u f \leq f \qquad \text{all } u > 0.$$

Then f is a potential in F and $E(f, f) \leq E(g, g)$. ///

Proof. Again we consider only the first alternative. Then the lemma follows from Lemma 1.1 and the estimate

$$\int dx \{f(x) - P_t f(x)\} f(x) \leq \int dx \{f(x) - P_t f(x)\} g(x)$$

$$= \int dx f(x) \{g(x) - P_t g(x)\}$$

$$\leq \int dx g(x) \{g(x) - P_t g(x)\} . ///$$

Remark. Lemma 3.3 is true without the restriction that g be in $\underline{F}$. For a proof we wait until Section 8 when we can apply results on random time change. We could prove it now using the technique of Lemma 1.7 if we knew that $\underline{F}$ contained at least one nontrivial potential. ///

3.2. Definition. The Radon measure μ has finite energy if there exists $c > 0$ such that

$$\int \mu(dx) f(x) \leq c \{E(f, f)\}^{\frac{1}{2}}$$

for f in $\underline{F}_{(e)} \cap C_{com}(\underline{X})$. The collection of all such measures is denoted by $\mathcal{M}$.

The collection of Borel $\varphi \geq 0$ such that $\varphi \cdot dx$ belongs to $\mathcal{M}$ is denoted by $\mathcal{M}^o$ ///

Clearly if μ belongs to $\mathcal{M}$ then there exists a unique potential, written $N\mu$, such that

$$(3.4) \qquad E(N\mu, g) = \int \mu(dx) g(x)$$

for g in $\underline{\underline{F}}_{(e)} \cap C_{com}(\underline{\underline{X}})$. We introduce the special notation

$$(3.5) \qquad E(\mu) = E(N\mu, N\mu)$$

and call $E(\mu)$ the energy of μ. Important compactness properties of $\mathcal{M}$ are summarized in

Lemma 3.4. *Let $\{\mu_n\}$ be a sequence in $\mathcal{M}$.*

(i) *If $N\mu_n$ converges weakly to f in $\underline{\underline{F}}_{(e)}$ then f is a potential and indeed $f = N\mu$ where μ is the vague limit of the μ_n.*

(ii) *If $E(\mu_n)$ is bounded and if $\mu_n \to \mu$ vaguely then μ is in $\mathcal{M}$ and $N\mu_n \to N\mu$ weakly in $\underline{\underline{F}}_{(e)}$.*

(iii) *$\mathcal{M}$ is complete relative to the energy metric E.* ///

This lemma is an immediate consequence of regularity. We omit the proof.

To make further progress we must validate (3.4) for general g in $\underline{\underline{F}}_{(e)}$ which means in particular that we must represent g by a refinement which is specified up to μ equivalence for every μ in $\mathcal{M}$. The main tool for this is a capacity associated with E.

3.3. Definition. For G an open subset of $\underline{\underline{X}}$ let

$$\mathrm{Cap}(G) = \inf E(f, f)$$

as f runs over the functions in $\underline{\underline{F}}_{(e)}$ such that $f \geq 1$ almost everywhere on G. If no such f exist let $\text{Cap}(G) = +\infty$. For A a general Borel subset of $\underline{\underline{X}}$ let

$$\text{Cap}(A) = \inf \text{Cap}(G)$$

as G runs over the open supersets of A. ///

3.4. Definition. A Borel set A is polar if $\text{Cap}(A) = 0$. A general set is polar if it is a subset of a Borel set which is polar. ///

Lemma 3.5. Let G be an open subset of $\underline{\underline{X}}$ such that $\text{Cap}(G) < +\infty$.

(i) There exists a unique function p^G in $\underline{\underline{F}}_{(e)}$ such that $E(p^G, p^G)$ is minimal among f in $\underline{\underline{F}}_{(e)}$ satisfying $f \geq 1$ almost everywhere on G.

(ii) $0 \leq p^G \leq 1$ almost everywhere and $p^G = 1$ almost everywhere on G.

(iii) p^G is a potential and indeed $p^G = N\nu$ with ν concentrated on $c\ell(G)$, the closure of G. ///

Proof. Let W be the subset of f in $\underline{\underline{F}}_{(e)}$ such that $f \geq 1$ almost everywhere on G. Clearly W is convex and closed and (i) follows directly. Concclusion (ii) follows upon noting that if f belongs to W then so does $\min(f,1)$ and $\max(f,0)$. To prove (iii) note first that if g is in $\underline{\underline{F}}_{(e)}$ and if $g \geq 0$ on G then $E(f+tg, f+tg) \geq E(f,f)$ for all $t > 0$ and therefore $E(f,g) \geq 0$. It only remains to adapt the proof of Lemma 3.1 considering restrictions to $c\ell(G)$ of $\underline{\underline{F}}_{(e)} \cap C_{com}(\underline{\underline{X}})$. ///

Proposition 3.6. For ν in $\mathcal{M}$ and for G an open subset of $\underline{\underline{X}}$

$$\nu(G) \leq \{E(\nu)\}^{\frac{1}{2}} \{\text{Cap}(G)\}^{\frac{1}{2}} . /// \tag{3.6}$$

If p^G were in $C_{com}(\underline{\underline{X}})$ this would follow from (3.4). In general this is false and instead we must approximate with the help of

Lemma 3.7. Let ν be in $\mathfrak{M}$.

(i) $(1-P_t)\ N\nu(x)$ belongs to $\mathfrak{M}^o$ for $t > 0$.

(ii) $\text{Lim}_{t \downarrow 0}\ (1/t)\ (1-P_t)\ N\nu(x)\ dx = \nu$ both vaguely and relative to the energy metric E. ///

Proof. For $T > 0$

$$\int_0^T ds\ P_s(1-P_t)N\nu = \int_0^t ds\ P_s N\nu - \int_T^{T+t} ds\ P_s N\nu\ .$$

Therefore $(1-P_t)\ N\nu$ is in $\mathfrak{M}^o$ and

$$G(1-P_t)\ N\nu \quad = \quad \int_0^t ds\ P_s N\nu\ .$$

and the lemma follows with the help of Lemma 3.4. ///

Now Proposition 3.6 follows from

$$\nu(G) \leq \text{Lim inf}_{t \downarrow 0}\ (1/t) \int_G dx(1-P_t)\ N\nu(x)$$

$$\leq \text{Lim inf}_{t \downarrow 0}\ (1/t) \int dx\ \ p^G(x)\ (1-P_t)\ N\nu(x)$$

$$= \text{Lim}_{t \downarrow 0}\ (1/t)\ E(p^G,\ G(1-P_t)N\nu)$$

$$= E(p^G, N\nu)$$

and the Cauchy-Schwarz inequality. ///

Corollary 3.8. If ν belongs to $\mathfrak{M}$ then ν charges no polar set. ///

Next we establish some properties of Cap which permit the application of Choquet's general theory.

Proposition 3.9. (i) $\mathrm{Cap}(G_1) \le \mathrm{Cap}(G_2)$ whenever $G_1 \subset G_2$ with both open.

(ii) If open $G_n \uparrow G$ then $\mathrm{Cap}(G_n) \uparrow \mathrm{Cap}(G)$. If Cap(G) is finite then also $p^{G_n} \to p^G$ in $\underline{\underline{F}}_{(e)}$ and $p^{G_n} \uparrow p^G$ almost everywhere.

(iii) (strong sub-additivity) For G_1, G_2 open

$$\mathrm{Cap}(G_1 \cup G_2) + \mathrm{Cap}(G_1 \cap G_2) \le \mathrm{Cap}(G_1) + \mathrm{Cap}(G_2). \;///$$

Proof. (i) is clear. To prove(ii) observe first that if $m < n$ then

$$g = \min(p^{G_m}, p^{G_n}) \ge 1$$

and therefore $E(g,g) \ge E(p^{G_m}, p^{G_m})$. But also

$$(3.7) \qquad E(p^{G_m}, p^{G_m}) \ge E(g,g) + E(p^{G_m} - g,\, p^{G_m} - g)$$

and we conclude that $g = p^{G_m}$ and so the p^{G_n} increase with n in the almost everywhere sense. If $\sup_n \mathrm{Cap}(G_n) < +\infty$ then the p^{G_m} converge weakly in $\underline{\underline{F}}_{(e)}$ to some f and it is easy to check that $f = p^G$. That also $p^{G_n} \to p^G$ strongly in $\underline{\underline{F}}_{(e)}$ follows directly from (3.7) with p^{G_m}, p^{G_n} for $m < n$ playing the roles of g and p^{G_m} respectively. This proves (ii). For (iii) we remark first that if G is open with $\mathrm{Cap}(G) < +\infty$ then $\mathrm{Cap}(G) = E(p^G, f)$ whenever f in $\underline{\underline{F}}_{(e)}$ satisfies $f = 1$ almost everywhere on G. This follows since

$$(3.8) \qquad E(p^G + t[f - p^G],\, p^G + t[f - p^G]) \ge E(p^G, p^G)$$

for all real t. It suffices to consider the case when $\mathrm{Cap}(G_1)$ and $\mathrm{Cap}(G_2)$

are finite and then

$$\mathrm{Cap}(G_1) + \mathrm{Cap}(G_2) - \mathrm{Cap}(G_1 \cup G_2) - \mathrm{Cap}(G_1 \cap G_2)$$

$$= E(p^{G_1 \cup G_2},\ p^{G_1} + p^{G_2} - p^{G_1 \cup G_2} - p^{G_1 \cap G_2})$$

which is ≥ 0 since $p^{G_1} + p^{G_2} - p^{G_1 \cap G_2} \geq 1$ almost everywhere on $G_1 \cup G_2$. (This follows from the validity of (3.8) for $t > 0$ with $p^{G_1 \cup G_2}$ playing the role of p^G and with $p^{G_1} + p^{G_2} - p^{G_1 \cap G_2}$ playing the role of f.) ///

We apply Choquet's theory of capacities. (See [35, chap. III. 2].)

<u>Theorem 3.10.</u> <u>For any Borel set A</u>

$$\mathrm{Cap}(A) = \sup \mathrm{Cap}(K)$$

<u>as K runs over the compact subsets of A.</u> ///

<u>Corollary 3.11.</u> <u>A Borel subset A is polar if and only if every compact subset of A is polar.</u> ///

<u>Corollary 3.12.</u> <u>If A is Borel and nonpolar, then there exists nontrivial μ in $\mathcal{M}$ such that</u> $\mu(\underline{X} - A) = 0$. ///

<u>Proof.</u> By Corollary 3.11 we can assume that A is compact. Let G_n be relatively compact open sets which decrease to A and let ν_n be as in Lemma 3.5 - (iii) with G_n playing the role of G. For each n clearly $\nu_n(\underline{X}) \geq \mathrm{Cap}(G_n) \geq \mathrm{Cap}(A)$ and it suffices to take μ any vague limit point of the ν_n. (Actually the proof of Proposition 3.9-(ii) shows that the entire sequence ν_n converges vaguely.) ///

Now we are ready to introduce refined versions.

3.5. Definition. A property is valid quasi-everywhere (abbreviated q.e.) if the exceptional set is polar. Two functions are quasi-equivalent if they differ only on a polar set, that is, if they are equal quasi-everywhere.///

3.6. Definition. $f_n \to f$ quasi-uniformly if there exists a decreasing sequence of open sets U_m with $\mathrm{Cap}(U_m) \downarrow 0$ such that $f_n \to f$ uniformly on $\underline{\underline{X}} - U_m$ for each m.///

3.7. A function f on $\underline{\underline{X}}$ is quasi-continuous on an open set G if there exists a decreasing sequence of open sets U_m with $\mathrm{Cap}(U_m) \downarrow 0$ such that f is defined and continuous on $G - U_m$ for each m. ///

Theorem 3.13. (i) Each f in $\underline{\underline{F}}_{(e)}$ has a representative uniquely specified up to to quasi-equivalence such that

3.8.1. f is quasi-continuous on $\underline{\underline{X}}$.

3.8.2. If $f_n \to f$ in $\underline{\underline{F}}_{(e)}$ then for a subsequence $f_n \to f$ quasi-uniformly.

(ii) If f in $\underline{\underline{F}}_{(e)}$ is represented by this quasi-continuous version then

$$(3.9) \quad \mathrm{Cap}\,\{x : (f(x)) > \varepsilon\} \leq (1/\varepsilon^2)\; E(f,f)$$

$$(3.10) \qquad E(f, N\mu) = \int \mu(dx)\, f(x)$$

for $\varepsilon > 0$ and for μ in $\mathcal{M}$.///

Proof. Uniqueness of the desired version is clear from 3.8.2. To establish existence fix f in $\underline{\underline{F}}_{(e)}$ and consider f_n in $\underline{\underline{F}}_{(e)} \cap C_{com}(\underline{\underline{X}})$ such that $f_n \to f$ in $\underline{\underline{F}}_{(e)}$. Since (3.9) is known to be valid for functions in $\underline{\underline{F}}_{(e)} \cap C_{com}(\underline{\underline{X}})$ we can assume after selecting a subsequence that 3.10

$$\mathrm{Cap}\{x : |f_n(x) - f_m(x)| > 1/N\} \leq 2^{-N}$$

for $m, n \geq N$. Then clearly the f_n converge quasi-uniformly to a quasi-continuous refinement of f. The relations (3.9) and (3.10) (with the help of (3.6)) extend routinely to these refinements and then 3.8.2 follows as above with the help of (3.9). ///

The statement of uniqueness in Theorem 3.13 - (i) will be improved in Lemma 3.16.

3.9. Convention. Unless otherwise specified every f in $\underline{\underline{F}}_{(e)}$ is represented by the version specified in Theorem 3.13.///

Corollary 3.14. (Maximum Principle.) Let $f = N\mu$ and $g = N\nu$ be potentials in $\underline{\underline{F}}_{(e)}$ with $g \geq f$ [a.e.μ]. Then $g \geq f$ quasi-everywhere.///

Proof. Let $h = \min(f, g)$ and observe that

$$E(h, f) = \int \mu(dx)h(x) = \int \mu(dx)f(x) = E(f, f)$$

and therefore

$$E(h, h) = E(f, f) + E(h-f, h-f)$$

But by Corollary 3.2 - (ii) h is a potential and therefore also

$$E(f, f) \geq E(h, h) + E(f-h, f-h)$$

which implies that $h = f$. ///

Applying Corollary 3.14 with $g = \min(f, c)$ we get

Corollary 3.15 Let $f = N\mu$ be a potential in $\underline{\underline{F}}_{(e)}$ such that $f \leq c$ [a.e. μ]. Then actually $f \leq c$ quasi-everywhere. ///

We finish this section with some technical results. The first was established by Fukushima in [22].

Lemma 3.16. Let f, g be quasi-continuous on an open subset G of $\underline{\underline{X}}$ and suppose that $f \geq g$ almost everywhere on G. Then also $f \geq g$ quasi-everywhere on G. ///

Proof. We will show that $\text{Cap}(A) = 0$ where $A = \{x \text{ in } G, f(x) < g(x)\}$. Fix $\varepsilon > 0$ and choose an open subset ω of G such that $\text{Cap}(\omega) < \varepsilon$ and such that f, g are continuous on $G-\omega$. Suppose first that ω has the following property : if x belongs to $G-\omega$ then every neighborhood U of x satisfies $\int_{U-\omega} dx > 0$. Then A must be contained in ω and we are done. In general it suffices to replace ω by the open set ω' of x in G having a neighborhood U_x such that $\int_{U_x-\omega} dx = 0$. ///

Lemma 3.17 Let ν be a bounded Radon measure such that $\nu(A) \leq c\ \text{Cap}(A)$ for A Borel and for fixed $c > 0$. Then ν is in $\mathcal{M}$. ///

Proof. It suffices to show that

$$\int \nu(dx)\ f(x) \leq \text{constant}$$

for $f \geq 0$ in $\underline{\underline{F}}_{(e)} \cap C_{com}(\underline{\underline{X}})$ such that $E(f,f) = 1$. But this follows since

$$\begin{aligned} &\int \nu(dx)\ f(x) \\ &\quad \leq \nu\{x: f(x) < 1\} + \Sigma_{k=0}^{\infty} 2^{k+1}\, \nu\{x : 2^k \leq f(x) < 2^{k+1}\} \\ &\quad \leq \nu(\underline{\underline{X}}) + c\ \Sigma_{k=0}^{\infty} 2^{k+1}\, \text{Cap}(\{x : f(x) \geq 2^k\}) \leq \nu(X) + c\, \Sigma_{k=0}^{\infty} 2^{k+1} 2^{-2k}. \end{aligned}$$

///

<u>Lemma 3.18.</u> <u>Every Radon measure ν which charges no polar set is the vague limit of an increasing sequence of measures ν_n in $\mathfrak{M}$.</u> ///

<u>Proof.</u> It suffices to consider the case when ν is bounded and because of Lemma 3.17 it suffices to establish the following result.

<u>3.10.1.</u> There exist Borel subsets X_N of $\underline{X}$ with $\nu(X_N) \downarrow 0$ such that $\nu(A) \leq N\,\mathrm{Cap}(A)$ for every Borel subset A of $\underline{X} - X_N$. ///

We first establish a preliminary result.

<u>3.10.2.</u> There exists $\alpha_N \downarrow 0$ such that for B Borel the inequality $\nu(B) \geq N\,\mathrm{Cap}(B)$ implies that $\nu(B) \leq \alpha_N$. ///

If 3.10.2 is false then there exist Borel B_n such that

$$\nu(B_n) \geq \alpha > 0 \ : \ \mathrm{Cap}(B_n) \leq 2^{-n-1} .$$

Then $\nu(\bigcap_{m=1}^{\infty} \bigcup_{n=m}^{\infty} B_n) \geq \alpha$ while for all m

$$\mathrm{Cap}\,(\bigcap_{m=1}^{\infty} \bigcup_{n=m}^{\infty} B_n) \leq \Sigma_{n=m}^{\infty} 2^{-n-1} = 2^{-m}$$

which contradicts our assumption that ν charges no polar set. To prove 3.10.1 we employ one of the standard techniques for proving the Radon Nikodym theorem. First let $\alpha = \inf\{N\ \mathrm{Cap}\,(A) - \nu(A)\}$ as A runs over the Borel subsets of $\underline{X}$. Clearly $-\nu(\underline{X}) \leq \alpha \leq 0$. Choose A_1 such that

$$N\ \mathrm{Cap}\,(A_1) - \nu(A_1) \leq \frac{1}{2}\,\alpha .$$

Then for $A \subset \underline{X} - A_1$ we have

$$\alpha \leq N\ \mathrm{Cap}(A \cup A_1) - \nu(A \cup A_1)$$

$$\leq N\ \mathrm{Cap}(A) + N\ \mathrm{Cap}(A_1) - \nu(A) - \nu(A_1)$$

and therefore $N\ \mathrm{Cap}(A) - \nu(A) \geq \frac{1}{2}\alpha$. Continuing in this way we find a sequence of disjoint Borel sets $A_1, A_2, \ldots$ such that $N\ \mathrm{Cap}(A_n) \leq \nu(A_n)$ for all n and such that $N\ \mathrm{Cap}(A) - \nu(A) \geq 2^{-n}\alpha$ for A a subset of $\underline{\underline{X}} - (A_1 \cup \cdots \cup A_n)$. Finally we let $X_N = \bigcup_{n=1}^{\infty} A_n$, we note that

$$\nu(X_N) = \Sigma\,\nu(A_n) \geq N\,\Sigma\ \mathrm{Cap}(A_n) \geq N\ \mathrm{Cap}(X_N)$$

and we apply 3.10.2 to conclude that $\nu(X_N) \downarrow 0$. ///

Corollary 3.19. Every Radon measure ν which charges no polar set is the vague limit of an increasing sequence of measures in $\mathcal{M}$ with bounded potentials. ///

Proof. It suffices to consider ν in $\mathcal{M}$ and then the corollary follows upon approximating ν by its restriction to the set where $N\nu \leq n$ and applying the maximum principle. ///

3.11. Extension. If ν is a Radon measure charging no polar set define

$$N\nu = \mathrm{Lim}_{n\uparrow\infty} N\nu_n \tag{3.11}$$

where ν_n is any sequence in $\mathcal{M}$ which increases to ν. This is independent of the choice of the approximating sequence since for μ in $\mathcal{M}$

$$\int \mu(dx)\, N\nu(x) = \mathrm{Lim} \int \mu(dx)\, N\nu_n(x) \tag{3.12}$$

$$= \mathrm{Lim} \int \nu_n(dx)\, N\mu(x) = \int \nu(dx)\, N\mu(x). \quad ///$$

The results of this section are applicable if we replace E by E_u for $u > 0$. This is so whether $(\underline{F}, E)$ is transient or recurrent. The corresponding extended Dirichlet space is $\underline{F}$ itself. We will use a subscript u for the associated objects. Thus for G open $\mathrm{Cap}_u(G) = \inf E_u(f,f)$ with the infimum taken as f runs over functions in $\underline{F}$ such that $f \geq 1$ almost everywhere on G. It is easy to see that the $\mathrm{Cap}_u, u > 0$ are all comparable and that $\mathcal{M}_u$, $u > 0$ is independent of $u > 0$.

We finish with some comparisions between E and E_1 in the case when $(\underline{F}, E)$ is transient.

Proposition 3.20. A set A is polar for E_1 if and only if it is polar for E. ///

Proof. Clearly $\mathrm{Cap}(A) \leq \mathrm{Cap}_1(A)$. Therefore by Corollary 3.11 it suffices to consider A compact such that $\mathrm{Cap}(A) = 0$ and to show that $\mathrm{Cap}_1(A) = 0$. Fix nonnegative g in $\underline{F}$ such that $g \geq 1$ almost everywhere on a neighborhood of A and choose a sequence of nonnegative f_n, $n \geq 1$ in $\underline{F}_{(e)}$ such that each $f_n \geq 1$ on a neighborhood of A (which depends on n) and such that $E(f_n, f_n) \to 0$. Possibly after selecting a subsequence we can assume also that $f_n \to 0$ almost everywhere. But then $g_n = \min(f_n, g)$ is in $\underline{F}$ and $g_n \to 0$ almost everywhere and $E_1(g_n, g_n)$ is bounded. After possibly replacing the g_n by Cesàro sums of a subsequence (see the paragraph following 1.6.1') we can assume that that $E_1(g_n, g_n) \to 0$ and therefore $\mathrm{Cap}_1(A) = 0$. ///

Before establishing the analagous result for quasi-continuity we show that quasi-continuity is a local property.

Lemma 3.21. If f is defined and quasi-continuous on a neighborhood of every point in an open set G, then f is quasi-continuous on G. ///

<u>Proof.</u> Note first that if G_1, G_2, U_1, U_2 are open and if f is continuous on $G_1 - U_1$ and on $G_2 - U_2$, then f is continuous on $(G_1 \cup G_2) - (U_1 \cup U_2)$. From this it follows that if G' is open with closure compact and contained in G, then for $\epsilon > 0$ there exists U' open such that $\mathrm{Cap}(U') < \epsilon$ and f is continuous on $G' - U'$. Choose G_n open such that each G_n has closure compact and contained in G and such that $G_n \uparrow G$. For $\epsilon > 0$ choose U_n open so that f is continuous on $G_n - U_n$ and $\mathrm{Cap}(U_n) < \epsilon\, 2^{-n}$ Then f is continuous on $G_n - U$ for all n and therefore f is continuous on $G - U$. Finally $\mathrm{Cap}(U) \leq \Sigma_n \mathrm{Cap}\,(U_n) < \epsilon$ and the lemma is proved. ///

<u>Proposition 3.22.</u> <u>Let G be an open subset of $\underline{X}$ and let f be specified up to quasi-equivalence on G. Then f is quasi-continuous relative to E_1 if and only if it is quasi-continuous relative to E.</u> ///

<u>Proof.</u> Because of Lemma 3.21 we can restrict attention to the case when G has compact closure. It suffices to consider a decreasing sequence of open subsets U_n of G such that $\mathrm{Cap}(U_n) \downarrow 0$ and show that also $\mathrm{Cap}_1(U_n) \downarrow 0$. But this can be proved in exactly the same way as Proposition 3.20. ///

<u>Remark.</u> Proposition 3.22 implies in particular that every f in $\underline{F}_{(e)}$ has a version which is quasi- continuous in the E_1 sense. This is also true when $(\underline{F}, E)$ is recurrent but again the proof must wait until Section 8. ///

4. Construction of Processes

In this section we begin with a regular Dirichlet space $(\underline{F}, E)$ on $L^2(dx)$ and construct a strong Markov process which is a Hunt process modulo a polar set. This result was first established by M. Fukushima [22] using a more indirect approach.

It will not matter in this section whether $(\underline{F}, E)$ is transient or recurrent.

We have already noted in Section 1 that the P_t are bounded operators (and indeed contractions) on $\underline{F}$. Thus for ν in $\mathcal{M}_1$ and for $t > 0$ there is a unique measure νP_t in $\mathcal{M}_1$ such that

$$(4.1) \qquad \int (\nu P_t)(dy)\, f(y) = \int \nu(dx)\, P_t f(x)$$

for f in $\underline{F}$. (Recall our convention that functions in $\underline{F}$ are represented by their quasi-continuous versions.) On the other hand it is easily established by the usual arguments that there exist subprobabilities $P_t(x, dy)$ defined for $t > 0$ and x in $\underline{X}$ and satisfying

4.1.1. $P_t(\cdot, A)$ is Borel measurable for $t > 0$ and for A a Borel subset of $\underline{X}$.

4.1.2. For $t > 0$ and for f in $\underline{F} \cap C_{com}(\underline{X})$

$$(4.2) \qquad P_t f(x) = \int P_t(x, dy)\, f(y)$$

for quasi-every x in $\underline{X}$. ///

From 4.1.2 and (4.1) it follows that

$$(4.3) \qquad \nu P_t(d\cdot) = \int \nu(dx)\, P_t(x, d\cdot)$$

for $t > 0$ and for ν in $\mathcal{M}_1$. Thus for f in $\underline{F}$

4.2

$$\int \nu(dx) \int P_t(x, dy) f(y)$$

$$= \int \nu P_t(dy) f(y)$$

$$= \int \nu(dx)\, P_t f(x)$$

and it follows from Fubini's theorem that (4.2) is valid [a.e. $\nu(dx)$] and therefore by Corollary 3.12 for quasi-every x. Thus 4.1.2 is valid for general f in $\underline{F}$. It follows that for $s, t > 0$ and for quasi-every x

(4.4) $$P_{s+t}(x, d\cdot) = \int P_s(x, dz)\, P_t(z, d\cdot)$$

Choose N_1 polar such that (4.4) is valid for $s, t > 0$ rational and for x in $\underline{X} - N_1$. By (4.3) and Corollary 3.12 the set of x such that $P_t(x, N_1) > 0$ for any rational $t > 0$ is polar. Thus there exists polar N_2 containing N_1 such that $P_t(x, N_1) = 0$ for $t > 0$ rational and for x in $\underline{X} - N_2$. Continuing in this way and taking the union we prove

Lemma 4.1. There exists a polar set N such that (4.4) is valid for rational $s, t > 0$ and for x in $\underline{X}$-N, and such that $P_t(x, N) = 0$ for x in $\underline{X}$-N and for rational $t > 0$. ///

For the moment we use as a preliminary sample space the set Ω_0 of all mappings ω from the nonnegative rationals into the augmented state space $\underline{X} \cup \{\partial\}$. Here ∂ is the usual "dead point" which we adjoint to $\underline{X}$ as an isolated point when $\underline{X}$ is compact and as the "point at infinity" otherwise. Functions f on $\underline{X}$ are automatically extended to $\underline{X} \cup \{\partial\}$ so that $f(\partial) = 0$. For $t \geq 0$ the past $\mathcal{F}_t$ is the sigma algebra generated by the coordinate variables $\omega(s)$, $s \leq t$. The Borel algebra $\mathcal{F}$ is the σ-algebra generated by all of the coordinate variables $\omega(s)$, $s \geq 0$. Standard arguments [10] establish the following two theorems.

Theorem 4.2. For x outside the exceptional set N of Lemma 4.1 there is a unique probability $\mathcal{P}_x$ on the Borel algebra $\mathcal{F}$ of Ω_0 such that

$$\mathcal{E}_x f_0[\omega(t_0)] \cdots f_n[\omega(t_n)]$$

$$= \int P_{t_1}(x, dy_1) \cdots \int P_{t_n - t_{n-1}}(y_{n-1}, dy_n)\, f_0(x) \cdots f_n(y_n)$$

for $0 = t_0 < \cdots < t_n$ all rational and for Borel $f_0, \ldots, f_n \geq 0$. ///

Of course $\mathcal{E}_x$ is the usual expectation functional which corresponds to the probability $\mathcal{P}_x$.

For $t \geq 0$ and rational let θ_t be the shift transformation defined on Ω_0 by

$$\theta_t\, \omega(s) = \omega(t+s).$$

Theorem 4.3. (Simple Markov Property.) For x outside the exceptional set N of Lemma 4.1, for $t \geq 0$ rational and for $\xi \geq 0$ a Borel function on Ω_0

$$\mathcal{E}_x(\theta_t \xi \mid \mathcal{F}_t) = \mathcal{E}_{\omega(t)}\, \xi \qquad [\text{a.e. } \mathcal{P}_x]. ///$$

Of course $\mathcal{E}_x(\,|\,)$ is the usual conditional expectation and $\theta_t \xi(\omega) = \xi(\theta_t \omega)$. Note that $\omega(t)$ avoids N and so $\mathcal{E}_{\omega(t)}\, \xi$ is well defined $[\text{a.e. } \mathcal{P}_x]$.

Our starting point for establishing regularity of sample paths is

Theorem 4.4 There exists a polar set N satisfying the conclusion of Lemma 4.1 such that for x in $\underset{\sim}{X} - N$ the probability $\mathcal{P}_x$ is concentrated on the set of trajectories ω having one sided limits $\omega(t \pm 0)$ for all real $t \geq 0$. ///

4.4

Proof. Let R_1 be a nonnegative random variable which is exponentially distributed with rate 1 (that is, $\mathcal{P}_x(R_1 > t) = e^{-t}$) and which is independent of the Borel algebra $\mathcal{J}$. The theorem will follow if we show that for each f in $\underline{F} \cap C_{com}(\underline{X})$ there exists a polar set N(f) such that for x in $\underline{X}$ - N(f) and for [a.e. $\mathcal{P}_x$] trajectory ω the composition $f[\omega(t)]$ has one sided limits for real $t < R_1$. (We take the usual liberty of assuming that all structure on Ω_0 has been transferred to an appropriate augmented sample space on which R_1 is also defined.) Fix one such f and choose f_n in $\underline{F}$ each having the form $f_n = G_1 \varphi_n$ with φ_n bounded and square integrable such that $f_n \to f$ strongly in $\underline{F}$ and quasi-uniformly. From the simple Markov property it follows that for $\varphi \geq 0$ and bounded and for quasi-every x the process $\{e^{-t} G_1 \varphi [\omega(t)], t \geq 0$ and rational$\}$ is a uniformly bounded supermartingale and so by standard estimates [35, Chap. VI] has one sided limits everywhere for [a.e. $\mathcal{P}_x$] trajectory ω. The same is true with $G_1\varphi$ replaced by f_n and therefore we need only show that for quasi-every x

(4.5) $\mathcal{P}_x\{\omega: \omega(t)$ is in U_m for some nonnegative rational $t < R_1\}$

decreases to 0 as $m \uparrow \infty$ where decreasing open sets U_m are chosen so that $Cap_1(U_m) \downarrow 0$ and such that $f_n \to f$ uniformly on the complement of each U_m. To show this fix S a finite set of nonnegative rationals and put

$$\sigma_m = \text{minimum}\ \{t \text{ in } S : \omega(t) \text{ is in } U_m\}$$

with the understanding that $\sigma_m = +\infty$ when not otherwise defined. The process $\{e^{-t} p_1^{U_m}[\omega(t)], t \text{ in } S\}$ is a supermartingale and σ_m is a stopping time so that [35,128] applies. By Lemma 3.16 we have $p_1^{U_m} = 1$ quasi-everywhere on U_m and so for quasi-every x

4.5

$$\mathcal{P}_x[\sigma_m < R_1] = \mathcal{E}_x I(\sigma_m < R_1)\, p_1^{U_m}[\omega(\sigma_m)]$$

$$= \mathcal{E}_x e^{-\sigma_m} p_1^{U_m}[\omega(\sigma_m)] \leq p_1^{U_m}(x)$$

which decreases to 0 as $m \uparrow \infty$. The theorem follows since this last estimate is independent of S. ///

At this point we restrict Ω_0 to the subset of trajectories ω such that one sided limits $\omega(t \pm 0)$ exist for real $t \geq 0$. Also for real $t \geq 0$ we define the trajectory variables

$$X_t(\omega) = \text{Lim}\ \omega(s)$$

with the limit taken as <u>rational</u> s <u>decrease</u> to t.

Since the U_m are open clearly (4.5) is for quasi-every x the same as

(4.5') $\mathcal{P}_x\{\omega\colon X_t \text{ in } U_m \text{ for some nonnegative real } t < R_1\}$.

Thus we can approximate general f in F by functions in $\underline{F} \cap C_{com}(\underline{X})$ and argue as in the proof of Theorem 4.4 to establish

<u>Proposition 4.5</u>. <u>For each f in $\underline{F}$ there exists a polar set N satisfying the conclusion of Theorem 4.4 such that for x in $\underline{X}$ - N and for [a.e. $\mathcal{P}_x$] trajectory ω the function f is defined and continuous on the range</u> $\{X_t(\omega),\ t \geq 0\}$. <u>In addition</u> $\text{Lim}_{t \uparrow \zeta} f(X_t) = 0$ <u>whenever</u> $X_{\zeta - 0} = \partial$ <u>for such</u> ω. ///

From now on $P_t f$ and $G_u f$ are understood to be defined by

$$P_t f(x) = \mathcal{E}_x f(X_t) \qquad t > 0 \tag{4.6}$$

$$G_u f(x) = \mathcal{E}_x \int_0^\infty dt\, e^{-ut} f(X_t) \qquad u \geq 0 \tag{4.6'}$$

whenever the right side makes sense. This is consistent with previous definitions. Moreover

Lemma 4.6. If f is in $L^2(dx)$ then $P_t f$ and $G_u f$ belong to $\underline{F}$ for $t, u > 0$. If $(\underline{F}, E)$ is transient then $P_t f$ is in $\underline{F}_{(e)}$ if f has a quasi-continuous refinement in $\underline{F}_{(e)}$ and $G\varphi$ is in $\underline{F}_{(e)}$ for φ in $\mathcal{M}^0$ ///

Note. Recall our Convention 3.9 that unless otherwise specified functions in $\underline{F}$ and $\underline{F}_{(e)}$ (in the transient case) are represented by their quasi-continuous versions. ///

To prove Lemma 4.6 note first that by the spectral theorem P_t and G_u are, except possibly for quasi-continuous refinements, bounded operators from $L^2(dx)$ to $\underline{F}$. For t rational and for bounded f in $\underline{F}$ the function $P_t f$ is in $\underline{F}$ by construction. This is also true for t real since by Proposition 4.5 the functions $P_s f \to P_t f$ quasi-everywhere as rational s decrease to t. Next fix $t > 0$ and bounded g in $\underline{F}$ such that $g > 0$ quasi-everywhere. The collection of nonnegative functions f for which $P_t \{\min(f, g)\}$ is in $\underline{F}$ is closed under pointwise limits and contains $\underline{F} \cap C_{com}(\underline{X})$ and therefore is the set of all nonnegative Borel functions. Finally for fixed nonnegative Borel f in $L^2(dx)$ the functions $P_t \min(f, ng) \to P_t f$ quasi-everywhere as $n \uparrow \infty$ and so $P_t f$ is in $\underline{F}$. This establishes the lemma for P_t and a similar argument works for $G_u, u > 0$ after first considering for bounded f in $\underline{\underline{F}}$ the approximation of $G_u f$ by Riemann sums in the $P_t f$. Finally the result for $\underline{\underline{F}}_{(e)}$ follow upon approximation from below by functions in $\underline{\underline{F}}$. ///

Corollary 4.7. Let A be dx null.

(i) For quasi-every x in $\underline{\underline{X}}$

$$\mathcal{P}_x[\sigma(\underline{\underline{X}} - A) = 0] = 1.$$

(ii) If $(\underline{\underline{F}}, E)$ is transient and if ν in $\mathcal{M}$ is nontrivial then $N\nu > 0$ quasi-everywhere.

(iii) If $(\underline{\underline{F}}, E)$ is recurrent and if ν in $\mathcal{M}_1$ is nontrivial then $\mathrm{Lim}_{u \downarrow 0} N_u \nu = +\infty$ quasi-everywhere. ///

Proof. It follows directly from Lemma 4.6 that for fixed $t > 0$

$$P_t(x, A) = 0 \qquad \text{q.e.}\, x \tag{4.7}$$

and for $u \geq 0$

$$G_u(x, A) = 0 \qquad \text{q.e.}\, x. \tag{4.7'}$$

Conclusion (i) follows from (4.7') since if $P_x[\sigma(\underline{\underline{X}} - A) > 0] > 0$, then $G_u(x, A) > 0$. To prove (ii) let $A = \{x : N\nu(x) = 0\}$. From the supermartingale property of $\{N\nu(X_t), t \geq 0\}$

$$P_t(x, \underline{\underline{X}} - A) = 0 \tag{4.8}$$

for quasi-every x in A. By irreducibility either A or $\underline{\underline{X}}$-$A$ is dx null and since ν is nontrivial, A is dx null. But then (4.7) and (4.8) together imply that A is polar. For (iii) we consider first the special case when $\nu = \varphi \cdot dx$ and let $A = \{x : G\varphi(x) < +\infty\}$. From the very definition of recurrence A is dx-null. The supermartingale property of $\{G\varphi(X_t), t \geq 0\}$ again gives (4.8) and we conclude as above that A is polar. For general ν it suffices by the above argument to show that $\mathrm{Lim}_{u \downarrow 0} N_u \nu = +\infty$ almost everywhere. But if this is not so then there exists nontrivial $\varphi \geq 0$ such that

$$\int \nu(dx)\, G\varphi(x) = \mathrm{Lim}_{u \downarrow 0} \int \nu(dx)\, G_u\varphi(x)$$

$$= \mathrm{Lim}_{u \downarrow 0} \int dx\, \varphi(x)\, N_u\nu(x)$$

is finite which contradicts the special case already established. ///

Remark. In general the exceptional sets in (4.7) and (4.7') depend on A in a nontrivial way. Also it is not true that for quasi-every x the measures $P_t(x, \cdot)$ charge no polar set. To see this it suffices to consider symmetrized compound Poisson processes with singular Lévy measures. Fukushima has shown in [23] that the ability to choose the exceptional set in (4.7) independent of t and A is equivalent to absolute continuity of the resolvent measures $G_u(x, \cdot)$ which in turn is equivalent to absolute continuity of the transition probabilities $P_t(x, \cdot)$. We do not know if in the general case it is possible to choose the exceptional set in (4.7) independent of t. (Of course it is trivial that the exceptional set in (4.7') is independent of u.) ///

We turn now to the strong Markov property.

4.2. Definition. A stopping time is a nonnegative Borel measurable function T on Ω_0 (possibly taking the value $+\infty$) such that for each $t > 0$ the subset $[T < t]$ of Ω_0 belongs to the past $\mathcal{F}_t$. The corresponding past $\mathcal{F}_T$ is the sigma algebra of Borel subsets Γ of Ω_0 such that for all $t > 0$ the intersection $\Gamma \cap [T < t]$ belongs to $\mathcal{F}_t$. The shift transformation θ_T is defined on the set $[T < \infty]$ by

$$\theta_T \omega(s) = \omega(T(\omega) + s). \ ///$$ 4.9

Theorem 4.8. Let T be a stopping time.

(i) The coordinate $X_T(\omega) = X_{T(\omega)}(\omega)$ is $\mathcal{F}_T$ measurable.

(ii) (Strong Markov property.) There exists a polar set N independent of T and satisfying the conclusion of Theorem 4.4 such that for x in $\underline{X}$-N and for $\xi \geq 0$ and Borel on Ω_0

$$\mathcal{E}_x(\theta_T \xi \mid F_T) = \mathcal{E}_{X(T)} \xi \ . ///$$

Proof. Fix a sequence of rational valued stopping times T_n which decrease to T as $n \uparrow \infty$. (For example take T_n to be the first positive number of the form $k/2^n$ which is $\geq$ T.) Then for K compact and for open $G_m \downarrow K$

$$[T < t] \cap [X_T \varepsilon K]$$

$$= \cap_{m=1}^{\infty} \cup_{N=1}^{\infty} \cap_{n=N}^{\infty} \{[T_n < t] \cap [X_{T_n} \varepsilon G_m]\}$$

and $\mathcal{F}_T$ measurability of X_T follows. To establish the strong Markov property choose a polar set N satisfying the conclusion of Theorem 4.4 and such that for x in $\underline{X}$-N and for [a.e $\mathcal{P}_x$] trajectory ω the functions

$$f_0(x), \ P_{t_1} f_1 P_{t_2 - t_1} \cdots P_{t_m - t_{m-1}} f_m(x)$$

are defined, bounded and continuous on the range $\{X_t(\omega), t \geq 0\}$ for every choice of $f_0, \ldots, f_m$ belonging to a fixed countable dense subset of $\underline{F}$ and of $0 < t_1 < \cdots < t_m$ rational. (Apply Proposition 4.5.) Finally put $\xi = f_0(X_0) \cdots f_m(X_{t_m})$, note that by the simple Markov property

$$\mathcal{E}_x\, I(A)\, I(T < \infty)\ \theta_{T_n}\ \xi \tag{4.10}$$

$$= \mathcal{E}_x I(A)\, I(T < \infty)\, f_0(X_{T_n})\, (P_{t_1} \cdots P_{t_m - t_{m-1}} f_m)(X_{T_n})$$

for all n and for A in the past $\mathcal{F}_T$ and pass to the limit $n \uparrow \infty$. ///

Remark. The strong Markov property extends in an obvious way to functions which are jointly measurable in the "past and future." We take this for granted below. ///

Theorem 4.9. (Quasi-left-continuity) There exists a polar set N satisfying the conclusion of Theorem 4.8-(ii) such that the following is true for all x in $\underline{X}$-N. If T_n, T are stopping times such that $T_n \uparrow T$ [a.e. $\mathcal{P}_x$], then $X_T = \text{Lim}_{n\uparrow\infty} X_{T_n}$ [a.e. $\mathcal{P}_x$] on the set $[T < +\infty]$. ///

Proof. We adapt the argument of Kunita and Watanabe [32]. Let R_1 be as in the proof of Theorem 4.4. Fix m < n and bounded f in $\underline{F}$. By Fubini's theorem and the strong Markov property

$$\mathcal{E}_x\Big(I(T_n < R_1) \int_{T_n}^{R_1} dt\, f(X_t) \;\Big|\; \mathcal{F}_{T_m}\Big)$$

$$= \mathcal{E}_x\big(I(T_n < R_1)\, G_1 f(X_{T_n}) \;\big|\; \mathcal{F}_{T_m}\big)$$

for x in $\underline{X}$-N with N as in Theorem 4.8-(ii). The analogous relation is true with T_n replaced by T. Applying the dominated convergence theorem for conditional expectations and applying Proposition 4.5 with $G_1 f$ playing the role of f we conclude that

4.11

$$\mathcal{E}_x(I(T < R_1)\, G_1 f(\mathrm{Lim}\, X_{T_n}) \mid \mathcal{F}_{T_m})$$

$$= \mathcal{E}_x(I(T < R_1)\, G_1 f(X_T) \mid \mathcal{F}_{T_m})$$

for x in X-N where N now depends on f. (Note that $P_x(T = R_1) = 0$ for quasi-every x.) But for bounded Borel g the function $g(\mathrm{Lim}\, X_{T_n})$ is measurable with respect to the sigma algebra generated by the union of the $\mathcal{F}_{T_m}$ and therefore

$$\mathcal{E}_x I(T < R_1)\, g\,(\mathrm{Lim}\, X_{T_n})\, G_1 f(\mathrm{Lim}\, X_{T_n})$$

$$= \mathcal{E}_x I(T < R_1)\, g(\mathrm{Lim}\, X_{T_n})\, G_1 f(X_T).$$

The theorem follows after approximating a countable dense set in $\underline{F} \cap C_{com}(\underline{X})$ as in Theorem 4.4 by functions $G_1 f$ as above. ///

The <u>death time</u> ζ is defined by

$$\zeta(\omega) = \inf\{t \geq 0 : X_t = \partial\}\ .$$

From the strong Markov property and the identity

$$P_\partial[X_t = \partial \text{ for } t \geq 0] = 1$$

it follows that modulo the usual exceptional set $X_t(\omega) = \partial$ for $t \geq \zeta$. Also from quasi-left continuity it follows that $X_{t-0} \neq \partial$ for $0 < t < \zeta$. Thus except for x in the polar set N of Theorem 4.9 the probabilities P_x are well defined on

<u>4.3</u>. <u>Standard Sample Space</u> Ω. This is the collection of maps ω from the half line $[0, \infty)$ into the augmented space $\underline{X} \cup \{\partial\}$ which satisfy the following

two conditions.

4.3.1. $\omega(\cdot)$ is right continuous and has one sided limits everywhere.

4.3.2. There exists a death time $\zeta(\omega)$ with $0 \leq \zeta(\omega) \leq +\infty$ such that $\omega(t) = \partial$ if and only if $t \geq \zeta(\omega)$ and such that $\omega(t-0) \neq \partial$ for $0 < t < \zeta$. ///

From now on all relevant structures introduced above are understood to be transferred to the standard sample space Ω.

Theorem 4.10. (Continuity of σ-algebras) Let N satisfy the conclusion of Theorem 4.9. Then the following is true for x in $\underline{X}$-N.

(i) If T_n, T are stopping times such that $T_n \downarrow T$ [a.e. $\mathscr{P}_x$], then for each event A in the intersection $\cap \mathscr{F}_{T_n}$ there exists A' in the past $\mathscr{F}_T$ such that A and A' differ only by a $\mathscr{P}_x$ null set.

(ii) If T_n, T are stopping times such that $T_n \uparrow T$ [a.e. $\mathscr{P}_x$], then for each event A is the past $\mathscr{F}_T$ there exists A' in the sigma algebra $\bigvee \mathscr{F}_{T_n}$ generated by the pasts $\mathscr{F}_{T_n}$ such that A and A' differ only by a $\mathscr{P}_x$ null set.///

Proof. In either case it suffices to show that

$$\mathscr{E}_x(\xi \mid \mathscr{F}_{T_n}) \to \mathscr{E}_x(\xi \mid \mathscr{F}_T) \tag{4.9}$$

in $L^1(dP_x)$ for a dense set of ξ in $L^1(dP_x)$. We begin with (i) and consider first the special case

$$\xi = \psi\, \theta_T \varphi \tag{4.10}$$

where ψ is bounded and $\mathscr{F}_T$ measurable and vanishes on $[T = \infty]$ and where

$$\varphi = f_0(X_{t_0}) \cdots f_m(X_{t_m}) \tag{4.11}$$

with the same understanding as in the proof of Theorem 4.8. Clearly

$$\mathcal{E}_{\mathbf{x}}(\psi I(T_n < +\infty)\, \theta_{T_n} \varphi \mid \mathcal{F}_{T_n})$$

$$= \psi I(T_n < +\infty)\, f_0(X_{T_n})\, P_{t_1} \cdots P_{t_m} f_m(X_{T_m})$$

converges [a.e. $\mathcal{P}_{\mathbf{x}}$] and therefore in $L^1(d\mathcal{P}_{\mathbf{x}})$ to

$$\mathcal{E}_{\mathbf{x}}(\xi \mid \mathcal{F}_T)$$

$$= \psi f_0(X_T)\, P_{t_1} \cdots P_{t_n} f_m(X_T)$$

and (4.9) follows since also $\psi I(T_n < +\infty)\, \theta_{T_n} \varphi$ converges to ξ. Thus (i) will be proved once we show that functions of the form (4.10) are dense in $L^1(d\mathcal{P}_{\mathbf{x}})$. But this follows easily from the observation that

$$(4.12)\quad I(t \le T) f(X_t) + \sum_{k=0}^{n-1} I(kt/n \le T < (k+1)t/n)\, \theta_T f(X_{t-kt/n})$$

converges to $f(X_t)$ for f in $C_{com}(\underline{\underline{X}})$ and $t > 0$. To prove (ii) we consider first the special case $\xi = \varphi$ with φ as in (4.11). There is no loss of generality in assuming that $T > 0$ and that $\mathcal{P}_{\mathbf{x}}[T = t_i] = 0$ for each i and then

$$\operatorname{Lim} \mathcal{E}_{\mathbf{x}}(I(t_i < T < t_{i+1})\, \xi \mid \mathcal{F}_{T_n})$$

$$= \operatorname{Lim} \mathcal{E}_{\mathbf{x}}(I(t_i < T_n < t_{i+1})\, \xi \mid \mathcal{F}_{T_n})$$

$$= \operatorname{Lim} I(t_i < T_n < t_{i+1}) \prod_{j=0}^{i} f_j(X_{t_j})\, \mathcal{E}_{\mathbf{x}}\Big(\prod_{j=i+1}^{m} f_j(X_{t_j}) \mid \mathcal{F}_{T_n}\Big)$$

at least in the $L^1(dP_{\mathbf{x}})$ sense. By the approximation (4.12) we need only consider convergence of

(4.13) $\mathscr{E}_x(\psi\,\theta_T\varphi' \mid \mathcal{F}_{T_n})$

on the set $[t_i < T < t_{i+1}]$ with φ' as in (4.10) and with ψ bounded and measurable with respect to $\bigvee \mathcal{F}_{T_n}$. But the limit of (4.13) is the same as the limit of

$$\psi\mathscr{E}_x(\theta_T\varphi' \mid \mathcal{F}_{T_n})$$

and after replacing $\theta_T\varphi'$ by $\theta_{T_n}\varphi'$ as above and applying quasi-left continuity, it is easy to see that the latter limit is $\psi\mathscr{E}_x(\theta_T\varphi' \mid \mathcal{F}_T)$ as required by (ii). (Note that the sets $[T \geq t]$ and $[kt/n \leq T < (k+1)\,t/n]$ actually belong to $\bigvee \mathcal{F}_{T_n}$.) ///

Conclusion (ii) states in the language of [35] that there are <u>no times of discontinuity</u>. Conclusion (i) implies in particlar that the sigma-algebras $\mathcal{F}_t$ are right continuous modulo $\mathscr{P}_x$ null sets for x in $\underline{X}$-N. We single out an important special case of this in

<u>Corollary 4.11.</u> (0-1 Law.) <u>Let A be a Borel subset of Ω which belongs to the infinitesimal future</u> $\mathcal{F}_{0+} = \bigcap_{\epsilon > 0} \mathcal{F}_+$ <u>and let N satisfy the conclusion of Theorem 4.9. Then</u>

(4.14) $\mathscr{P}_x(A) = 0$ or 1 for x in $\underline{X}$-N. ///

<u>Proof.</u> By Theorem 4.10 the set A belongs to $\mathcal{F}_0$ modulo a $\mathscr{P}_x$ null set, and (4.14) follows immediately. ///

4.15

Following [35] (except that we imitate the recent literature and replace the adjective "accessible" by "predictable") we introduce

4.4. Definition. A stopping time T is $\mathcal{P}_x$ predictable if there exists a sequence of stopping times T_n such that

(4.15) $$\mathcal{P}_x [T_n \uparrow T] = 1$$

(4.16) $$\mathcal{P}_x [T_n < T] = 1 \qquad \text{for all } n.$$

A stopping time T is $\mathcal{P}_x$ totally unpredictable if for any sequence of stopping times T_n satisfying (4.15)

(4.17) $$\operatorname{Lim}_{n \uparrow \infty} \mathcal{P}_x [T_n < T < +\infty] = 0. \ ///$$

The next theorem characterizes predictability and shows among other things that modulo polar sets the adjective $\mathcal{P}_x$ can be dispensed with in Definition 4.4.

Theorem 4.12 . There exists N satisfying the conclusion of Theorem 4.9 such that the following is true for x in $\underset{\sim}{X}$-N.

(i) A stopping time T is $\mathcal{P}_x$ predictable if and only if

(4.18) $$\mathcal{P}_x [T < +\infty ; \ X_T \neq X_{T-0}] = 0.$$

(ii) A stopping time T is $\mathcal{P}_x$ totally unpredictable if and only if

(4.19) $$\mathcal{P}_x [\, T < +\infty ; X_T = X_{T-0}] = 0. \ ///$$

Proof. It follows from quasi-left continuity that (4.18) is true whenever T is $\mathcal{P}_x$ predictable. By T44 and T45 in [35, Chap. VII] we need only show that if (4.18) is true then T is predictable. By T46 in [35, Chap. VII] we need only show that there exists no uniformly integrable right continuous martingale $\{Y_t\}$

with a discontinuity at the time T. By the maximal inequality for such martingales we need only consider the special case where Y_t is a right continuous version of $\mathcal{E}_x(\varphi | \mathcal{F}_t)$ with φ given by (4.11). After possibly expanding N we can assume further that each f_i is represented

$$(4.20) \qquad f_i(x) = \mathcal{E}_x \int_0^\infty dt\, e^{-t} g_i(X_t) \qquad x \text{ in } \underline{\underline{X}}\text{-N}$$

with g_i bounded and integrable, the point being that in this case $P_s f_i \to f_i$ as $s \downarrow 0$ and also $P_s f_i \to P_t f_i$ as $s \to t$ with the convergence being uniform on $\underline{\underline{X}}$-N. But then

$$Y_t = I(t_m \le t)\ \Pi_{j=0}^m\ f_j(X_{t_j})$$

$$+ \Sigma_{i=0}^{m-1}\ I(t_i \le t < t_{i+1})\ \Pi_{j=0}^i\ f_j(X_{t_j})\ P_{t_{i+1}-t} f_{j+1} \cdots P_{t_m - t_{m-1}} f_m(X_t)$$

can have discontinuities only when the trajectory $X_.$ has discontinuities and the theorem is proved.///

We turn now to the hitting times

$$\sigma(A) = \inf \{ t \ge 0 :\ X_t \text{ is in } A \}$$

defined for Borel subsets A of $\underline{\underline{X}}$ with the understanding that $\sigma(A) = +\infty$ if not otherwise defined. We begin with open sets.

<u>Theorem 4.13.</u> <u>Let G be an open subset of the augmented state space</u> $\underline{\underline{X}} \cup \{\partial\}$.

(i) <u>$\sigma(G)$ is a stopping time.</u>

(ii) <u>If G has finite 1-capacity then for any $u > 0$</u>

(4.21) $\mathcal{E}_x \exp\{-u\,\sigma(G)\} = p_u^G(x)$ quasi-everywhere.

(iii) The left side of (4.21) is always quasi-continuous for $u \geq 0$.///

Proof. (i) is standard. (See for example [2, p. 54].) For (ii) let h be the minimum of the two sides of (4.21). At this stage we cannot be sure that h is quasi-continuous. Except for this it follows from Lemma 3.3 that h is a u-potential in $\underline{F}$ which is dominated in E_u norm by p_u^G. But $h \geq 1$ almost everywhere (and indeed quasi-everyhere) on G and so by the very definition of p_u^G the E_u norm of h dominates the E_u norm of p_u^G and the proof of Corollary 3.2 shows that $h = p_u^G$ in $\underline{F}$. Thus the left side of (4.21) dominates the right side almost everywhere. The opposite inequality follows as in the proof of Theorem 4.4. Finally equality quasi-everywhere in (4.21) follows with the help of Lemma 4.6 since clearly $P_t\mathcal{E}_{\cdot} \exp\{-u\sigma(G)\} \to \mathcal{E}_{\cdot} \exp\{-u\sigma(G)\}$ quasi - everywhere as $t \downarrow 0$. Conclusion (iii) follows upon considering the infimum of the left side of (4.21) with $p_u^{G'}$ (with $p_1^{G'}$ when u=0) where G' has finite 1-capacity and then letting G' increase to $\underline{X}$.///

Theorem 4.14. (Measurability of hitting times.) Let A be a Borel subset of $\underline{X}$.

(i) There exist stopping times $\sigma_i(A)$, $\sigma_e(A)$ and a polar set N depending on A such that $\sigma_e(A) \leq \sigma(A) \leq \sigma_i(A)$ everywhere on Ω and such that

$$P_x[\sigma_e(A) = \sigma_i(A)] = 1$$

for x in $\underline{X}$ - N.

(ii) Possible choices for $\sigma_i(A)$ and $\sigma_e(A)$ are

4.18

$$\sigma_i(A) = \text{Lim } \sigma^*(K_n)$$

$$\sigma_e(A) = \text{Lim } \sigma(G_n)$$

where the K_n form a particular increasing sequence of compact subsets of A and the G_n form a particlar decreasing sequence of open supersets of A.

(iii) *For $u \geq 0$ the function*

(4.22) $$h(x) = \mathcal{E}_x \exp\{-u\,\sigma(A)\}$$

is quasi-continuous. ///

Proof. Consider first A compact, let G_n be any decreasing sequence of open sets whose intersection is A and define $\sigma_e(A) = \text{Lim } \sigma(G_n)$. By quasi-left continuity there exists a polar set N such that $P_x[\sigma_e(A) = \sigma(A)] = 1$ for x in $\underline{X}$ - N. Since the set $[\sigma_e(A) = \sigma(A)]$ is Borel (see the proof of Theorem 4.8-(i)) we can satisfy (i) and (ii) for this special case by letting $\sigma_i(A) = \sigma_e(A)$ whenever $\sigma_e(A) = \sigma(A)$ and $\sigma^i(A) = +\infty$ otherwise. Conclusion (iii) follows directly from the corresponding result for open sets. To handle general Borel A we imitate Hunt's original argument and apply the Choquet extension theorem. Fix a bounded measure μ charging no polar set and define $\lambda(A)$ for A open or compact by

(4.23) $$\lambda(A) = \int \mu(dx)\ \mathcal{E}_x \exp\{-\sigma(A)\} .$$

We want to conclude that for any Borel set A

(4.24) $$\text{supremum } \lambda(K) = \text{infimum } \lambda(G)$$

as K runs over the compact subsets of A and as G runs over the open supersets of A. For this we must check that λ satisfies that following conditions, (See [35, Chap. III. 2].)

4.19

4.5.1. $\lambda(G_1) \leq \lambda(G_2)$ whenever $G_1 \subset G_2$ with G_1, G_2 open.

4.5.2. For K compact, $\lambda(K)$ = infimum $\lambda(G)$ as G runs over the open supersets of K.

4.5.3. $\lambda(G_1 \cup G_2) + \lambda(G_1 \cap G_2) \leq \lambda(G_1) + \lambda(G_2)$ for G_1, G_2 open. ///

Property 4.5.1 is clear and 4.5.2 follows from the first part of the proof. Property 4.5.3 follows from

$$\mathcal{E}_x \exp\{-\sigma(G)\} = \mathcal{P}_x[\sigma(G) < R_1]$$

and from

$$I(\sigma(G_1 \cup G_2) < R_1) + I(\sigma(G_1 \cap G_2) < R_1)$$

$$\leq I(\sigma(G_1) < R_1) + I(\sigma(G_2) < R_2).$$

Thus (4.24) is established. To complete the proof fix a Borel set A and take μ in (4.23) equivalent to dx. Then there exist increasing compact subsets K_n of A and decreasing open supersets G_n of A such that $\mathcal{P}_x[\sigma_e(A) = \sigma_i(A)]=$ for almost every x. (Of course $\sigma_i(A)$ and $\sigma_e(A)$ are as is conclusion (ii). But again the functions

$$h_e(A) = \mathcal{E}_x \exp\{-\sigma_e(A)\}; \; h_i(x) = \mathcal{E}_x \exp\{-\sigma_i(A)\}$$

are quasi-continuous. Thus by Lemma 3.16 $h_e = h_i$ quasi-everywhere and (ii) is proved. Again (iii) follows from the corresponding results for open sets.///

Remark 1. To our knowledge the function (4.22) need not be Borel measurable although it is quasi-equivalent to a Borel measurable function. For this reason other authors work with the wider notion of" nearly Borel" measurability. We prefer not to do this. Notice that conclusion (iii) has the following important consequence. For A Borel define

$$\sigma^{+}(A) = \inf \{t > 0 : X_t \text{ is in } A\}.$$

Then $\mathcal{P}_x[\sigma(A) = \sigma^{+}(A)] = 1$ for quasi-every x. This is true since as $t \downarrow 0$

$$P_t \mathcal{E}_{\cdot} \exp\{-\sigma(A)\}(x) \to \mathcal{E}_x \exp\{-\sigma^{+}(A)\}$$

quasi-everywhere by the simple Markov property and for a sequence

$$P_t \mathcal{E}_{\cdot} \exp\{-\sigma(A)\}(x) \to \mathcal{E}_x \exp\{-\sigma(A)\}$$

quasi-everywhere by quasi-continuity of the limiting function. (See the last part of the proof of Theorem 4.13.) The exceptional x set depends on A and, to our knowledge, need not be Borel measurable. ///

Remark 2. The proof of Theorem 4.14 shows that if $\text{Cap}_u(A) < +\infty$ then (4.22) is a u-potential. This is also true for $u=0$ when $(\underline{F}, E)$ is transient. For the proof it suffices to consider first A with compact closure so that $\text{Cap}_u(A)$ is bounded for $0 \leq u < 1$ and then pass to the limit in A. ///

Proposition 4.15. *If N is polar then*

$$\mathcal{P}_x[\sigma(N) = +\infty] = 1 \tag{4.25}$$

for quasi-every x. Conversely if N is Borel and if (4.25) is valid almost everywhere then N is polar. ///

Proof. The direct part follows immediately from Theorem 4.14-(ii) and from (4.21). The converse follows from Theorem 4.14 - (ii) and Lemma 3.16. ///

Proposition 4.16. *Assume that $(\underline{F}, E)$ is recurrent and let M be a nonpolar subset of $\underline{X}$. Then for quasi-every x*

(4.26) $$\mathcal{P}_x[\ \theta_n\ \sigma(M) < +\ \infty \text{ for all } n\ ] = 1.\ ///$$

Proof. By the proof of Theorem 4.14 - (iii) the function $h(x) = \mathcal{P}_x[\sigma(M) < +\ \infty]$ is quasi-continuous. For $t > 0$ the difference $h - P_t\ h$ is nonnegative and as in the proof of Theorem 1.6

$$\int_0^T ds\ P_s\ (h-P_t h\ )$$

remains bounded as $T \uparrow \infty$. Thus $h = P_t h$ almost everywhere and therefore quasi-everywhere and in particular $h(X_t)$, $t \geq 0$ is a martingale relative to the $\mathcal{P}_x$. It follows in particular that the set $A = \{x : h(x) = 1\}$ is irreducible in the sense of Section 1. Since A contains M it is nonpolar. By Corollary 4.7-(i) it is not dx null and so by irreducibility $\underline{\underline{X}}$ - A is dx null. Thus $h = 1$ almost everywhere and therefore quasi-everywhere and the proposition follows since also $P_n h = 1$ quasi-everywhere ///

We apply Proposition 4.15 and reason as in the paragraph preceeding Lemma 4.1 to prove

Theorem 4.17. There exists a Borel polar set N satisfying the conclusion of Theorem 4.12 such that $\mathcal{P}_x[\sigma(N) = +\ \infty] = 1$ for x in $\underline{\underline{X}}$ - N.///

Theorems 4.8, 4.9, 4.10 and 4.17 together give in the terminology of [2]

Theorem 4.18: There exists a Borel polar set N such that $\{\mathcal{P}_x : x \text{ in } \underline{\underline{X}}\text{-}N\}$ is a Hunt process on the state space $\underline{\underline{X}}$ - N. ///

4.6. Convention. In the remainder of the book we will generally ignore polar sets. Functions specified up to quasi-equivalence will be treated as if they are defined everywhere. Relations valid quasi-everywhere will be stated as if valid identically.///

4.22

We close this section with some additional notations which will be used throughout the book. For A a Borel subset of $\underset{=}{X}$ the hitting operators are defined by

(4.27) $$H^A f(x) = \mathcal{E}_x [\sigma(A) < +\infty;\ f(X_{\sigma(A)})]$$

$$H_u^A f(x) = \mathcal{E}_x e^{-u\sigma(A)} f(X_{\sigma(A)}) \qquad u > 0.$$

The last exit time is defined by

(4.28) $$\sigma^*(A) = \sup \{t > 0 : X_{t-0} \text{ is in } A\}$$

with the understanding that $\sigma^*(A) = -\infty$ when not otherwise defined. It follows from Theorem 4.14 that (4.27) is well defined at least for $f \geq 0$ and Borel. The last exit time $\sigma^*(A)$ is clearly Borel measurable for A open. General Borel A can always be handled by combining Theorem 4.14 with results on time reversal established in Section 5.

5. An Approximate Markov Process

In this section $(\underline{F}, E)$ is a transient regular Dirichlet space.

5.1. Convention. D_k, $k \geq 1$ is an increasing sequence of open subsets of $\underline{X}$ such that $D_k \uparrow \underline{X}$ and such that each D_k has compact closure. The complements $\underline{X} - D_k$ are denoted by M_k. Often k and $\sim k$ will be used in place of D_k and M_k for subscripts and superscripts.///

By the second remark following Theorem 4.14 the functions $H^k 1$ are potentials and so there exists unique measures L_k in $\mathcal{m}$ such that

$$(5.1) \qquad H^k 1 = NL_k .$$

We begin by establishing a result on time reversal for the measures

$$(5.2) \qquad \mathcal{P}^{(k)} = \int L_k(dx)\, \mathcal{P}_x .$$

The time reversal operator ρ_k and the truncation operator τ_k are defined on $\Omega \cap [\sigma(D_k) < \infty]$ by

$$\rho_k \omega(t) = \begin{cases} \omega(\sigma^*(D_k) - t - 0) & 0 \leq t < \sigma^*(D_k) \\ \partial & t \geq \sigma^*(D_k) \end{cases}$$

$$\tau_k \omega(t) = \begin{cases} \omega(t) & 0 \leq t < \sigma^*(D_k) \\ \partial & t \geq \sigma^*(D_k). \end{cases}$$

Note that by transience

$$\mathcal{P}_x [\sigma(D_k) < +\infty, \ \sigma^*(D_k) = +\infty] = 0$$

for quasi-every x.

Lemma 5.1. For $\xi \geq 0$ on Ω and for $k, \ell \geq 1$

(5.3) $\mathcal{E}^{(k)}[\sigma(D_\ell) < +\infty;\ \xi\circ\rho_\ell] = \mathcal{E}^{(\ell)}[\sigma(D_k) < +\infty;\ \xi\circ\tau_k]$ ///

Proof. We begin by establishing

(5.4) $$\mathcal{E}_x[t_n < \sigma^*(D_\ell) < +\infty;\ f_0(X_{\sigma^*(D_\ell)-0})\cdots f_n(X_{\sigma^*(D_\ell)-t_n-0})]$$

$$= N\{L_\ell \cdot f_0 P_{t_1}\cdot f_1 \cdots P_{t_n - t_{n-1}}\cdot f_n\}(x)$$

for $0 < t_1 < \cdots < t_n$, for bounded $f_0, \ldots, f_n \geq 0$ on $\underline{X}$ and for quasi-every x. The transition operators on the right side of (5.4) are understood to be acting on measures in $\mathfrak{m}$ as in (4.1). It suffices to establish (5.4) almost everywhere since the right side is a potential and since the left side can be recovered quasi-everywere upon application of P_t and passage to the limit $t \downarrow 0$. (See Lemma 4.6.) Therefore we can replace (5.4) by

(5.4′) $$\int dx\, \varphi(x)\, \mathcal{E}_x[t_n < \sigma^*(D_\ell) < +\infty;\ f_0(X_{\sigma^*(D_\ell)-0})\cdots f_n(X_{\sigma^*(D_\ell)-t_n-0})]$$

$$= \int L_\ell(dy)\, f_0(y)\, P_{t_1}\cdots P_{t_n-t_{n-1}}\{f_n\, G\varphi\}(y)$$

for $\varphi \geq 0$ in $L^1(dx)$ such that both φ and $G\varphi$ are bounded. Also we can assume that $f_0, \ldots, f_n$ are in $\underline{F} \cap C_{com}(\underline{X})$. The left side of (5.4′)

$$= \mathrm{Lim}_{p\uparrow\infty} \Sigma_{k=0}^{\infty} \int dx\, \varphi(x)\, P_{k/p}\, f_n \cdots P_{t_1}\, f_0 (1 - P_{1/p})\, H^\ell 1(x)$$

$$= \mathrm{Lim}_{p\uparrow\infty} \Sigma_{k=0}^{\infty} \int L_\ell(dy) \int_0^{1/p} ds\, P_s\, f_0\, P_{t_1}\cdots f_n\, P_{k/p}\, \varphi(y).$$

Fix $q > 0$ and let $p\uparrow\infty$ through a sequence of multiples of q. By the spectral theory applied directly to the Hilbert space $\underline{F}_{(e)}$

$$(1/p)\ \Sigma_{k=p/q}^{\infty}\ P_{k/p}\ \varphi \rightarrow GP_{1/q}\ \varphi$$

in $\underline{\underline{F}}_{(e)}$ and since $p\int_0^{1/p} ds\ P_s$ converges strongly to the identity as an operator on $\underline{\underline{F}}_{(e)}$ the left side of (5.4′) differs from

$$\int L_\ell(dy)\ f_0(y)\ P_{t_1} \cdots P_{t_n - t_{n-1}} \{f_n\ GP_{1/q}\varphi\}(y)$$

by a term which goes to zero as $q \uparrow \infty$ and (5.4′) follows. In (5.3) it suffices to consider $\xi = f_0(X_0) \cdots f_n(X_{t_n})$ and then (5.3) follows from

$$\int L_k(dx)\ \mathcal{E}_x[\sigma(D_\ell) < +\infty;\ \xi \circ \rho_\ell]$$

$$= \int L_k(dx)\ \mathcal{E}_x[t_n < \sigma^*(D_\ell) < +\infty;\ f_0(X_{\sigma^*(D_\ell)-0}) \cdots f_n(X_{\sigma^*(D_\ell)-t_n-0})]$$

$$= \int L_k(dx)\ N\ \{L_\ell \cdot f_0\ P_{t_1} \cdots P_{t_n - t_{n-1}} \cdot f_n\}(x)$$

$$= \int L_\ell(dy)\ f_0(y)\ P_{t_1} \cdots P_{t_n - t_{n-1}}\ f_n\ H^k\ 1(y). ///$$

Now we are ready to adapt Hunt's construction of "approximate Markov chains" as outlined in [26]. This was first done in continuous time by M. Weil [51].

For each k let Ω_k be the subcollection of ω in Ω satisfying

5.2. $\omega(0) = \partial$ or $\omega(0)$ is in the closure $c\ell(D_k)$. ///

There is a unique trajectory ω in Ω_k such that $\omega(t) = \partial$ for $0 \leq t \leq +\infty$. We refer to this trajectory as the <u>dead trajectory</u> and denote it by δ_k. We consider Ω_k with the Skorohod metric as defined for a special case and for compact time intervals in [39, Chap. VII]. A simple extension of the results in [39] shows that relative to the Skorohod topology Ω_k is a complete separable metric space. The mapping J_k from Ω_{k+1} to Ω_k is defined by

$$J_k\omega(t) = \begin{cases} \partial \text{ for all } t \text{ if } \sigma(D_k) = +\infty \\ \omega(\sigma(D_k)+t) \text{ if } \sigma(D_k) < +\infty . \end{cases}$$

Clearly each J_k is Borel measurable and surjective. The inverse limit of the Ω_k is the collection Ω_∞^0 of sequences $\{\omega_k\}_{k=1}^\infty$ with each ω_k in Ω_k and such that $J_k\omega_{k+1} = \omega_k$ for all k. The extended sample space is the reduced inverse limit $\Omega_\infty = \Omega_\infty^0 - \{\delta\}$ where δ is the dead sequence in Ω_∞^0 whose components are the dead trajectories δ_k. We denote by $J_{\infty,k}$ the natural projection of Ω_∞ onto Ω_k. It follows from [39, Chap. V] that Ω_∞ is a separable metric space and an absolute Borel set in the product Skorohod topology and that the projections $J_{\infty,k}$ are Borel measurable. The point of all this is

Theorem 5.2. There exists a unique countably additive measure $\mathcal{P}$ on the extended sample space Ω_∞ such that

(5.5) $$\mathcal{E}\,\xi \circ J_{\infty,k} = \mathcal{E}^{(k)}\,\xi$$

for each k and for $\xi \geq 0$ on Ω_k and vanishing on δ_k. ///

Proof. Note first that for $\varphi \geq 0$ on $\underline{X}$

$$\mathcal{E}^{(k+1)}[\sigma(D_k) < +\infty;\ \varphi(X_{\sigma(D_k)})]$$

$$= \mathcal{E}^{(k)}[\varphi(X_{\sigma^*(D_k)})]$$

$$= \mathcal{E}^{(k)}\,\varphi(X_0)$$

where we have applied Lemma 5.1 twice. From this and the strong Markov property follows

$$\mathscr{P}^{(k+1)}\xi \circ J_k = \mathscr{P}^{(k)}\xi$$

which is the necessary consistency condition for the existence of a finitely additive measure satisfying (5.5). To establish countable additivity it suffices to show that if A_k is a Borel subset of $\Omega_k - \{\delta_k\}$ and if the inverse images $J^{-1}_{\infty,k}(A_k)$ in Ω_∞ decrease with k and if

$$\mathscr{P}^{(k)}(A_k) \geq a > 0$$

for all k then

(5.6) $$\bigcap_{k=1}^{\infty} J^{-1}_{\infty,k}(A_k) \text{ is nonempty.}$$

According to [39, Theorem 3.2, p.139] there exists a sequence of compact metric spaces Ω^*_k and for each k a surjective continuous map $J^*_k : \Omega^*_{k+1} \to \Omega^*_k$ and an injective Borel map $\Phi_k : \Omega_k \to \Omega^*_k$ such that $J^*_k \Phi_{k+1} = \Phi_k J_k$. The inverse limit Ω^*_∞ of the Ω^*_k is a compact metric space and the projections $J^*_{\infty,k} : \Omega^*_\infty \to \Omega^*_k$ are continuous. Let $\mathscr{P}^{(k)*}$ be the unique Borel measure on Ω^*_k such that

$$\mathscr{P}^{(k)*}(B^*) = \mathscr{P}^{(k)}(\Phi_k^{-1} B^*)$$

for B^* a Borel subset of Ω^*_k. By Kuratowski's theorem [39, Theorem 3.9, p.21] the images $A^*_k = \Phi(A_k)$ are Borel subsets of Ω^*_k and of course $\mathscr{P}^{(k)*}(A^*_k) = \mathscr{P}^{(k)}(A_k)$. Choose compact subsets B^*_k of A^*_k such that

$$P^{(k)*}(B^*_k) \geq \mathscr{P}^{(k)*}(A^*_k) - a2^{-k}$$

and define

$$C_2^* = B_2^* \cap J_1^{*-1} (B_1^*)$$

$$C_{k+1}^* = B_{k+1}^* \cap J_k^{*-1} (C_k^*) \qquad k \geq 2.$$

Each C_k^* is compact in Ω_k^* and, what is really the point, the inverse images $J_{\infty,k}^{*-1} (C_k^*)$ are compact and of course decreasing in the inverse limit Ω_∞^*. Clearly $(J_1^*)^{-1} (A_1^*)$ contains A_2^* and so

$$\mathcal{P}^{(2)*}(C_2^*) \geq \mathcal{P}^{(2)*} (B_2^*) - \mathcal{P}^{(1)*} (A_1^* - B_1^*)$$

$$\geq a - a/4 - a/2$$

and similarly for $k \geq 2$

$$\mathcal{P}^{(k+1)*} (C_{k+1}^*) \geq \mathcal{P}^{(k+1)*}(B_{k+1}^*) - \mathcal{P}^{(k)*} (A_k^* - C_k^*)$$

$$\geq a - a/2^{k+1} - (a/2 + \cdots + a/2^k).$$

Thus C_k^* is nonempty for each k and therefore $\bigcap_{k=1}^\infty J_{\infty,k}^{*-1} (C_k^*)$ is nonempty. Now (5.6) follows and the theorem is proved.///

Next we introduce trajectory variables parametrized by an artifical two sided time scale for $\omega = \{\omega_k\}_{k=1}^\infty$ in Ω_∞. Let k_0 be the first integer k such that $\omega_k \neq \delta_k$ and for $t \geq 0$ define

$$X_t(\omega) = \omega_{k_0}(t).$$

For $t < 0$ there is at most one integer $k_t > k_0$ such that

(5.7) $$\sigma(D_{k_0}, \omega_{k_t}) \geq |t|$$

$$\sigma(D_{k_0}, \omega_{k_t - 1}) < |t| .$$

Define

$$X_t(\omega) = \omega_{k_t}(\sigma(D_{k_0}) + t) \tag{5.7}$$

if k_t exists and otherwise define $X_t(\omega) = \partial$. (Actually k_t can be replaced by any larger integer in (5.7) without changing the result. The idea is that $\sigma(D_{k_0})$ is the "zero point" of the time scale.) The coordinates X_t are Borel measurable on Ω_∞ and generate the Borel algebra on Ω_∞. Indeed ω in Ω_∞ is determined by its coordinates $X_t(\omega)$. The X_t do not in general form a Markov process relative to $\mathcal{P}$. First hitting times $\sigma(A)$, last exit times $\sigma^*(A)$, the death time ζ and the birth time ζ^* are defined in the obvious way. The time reversal operator ρ is defined so that

$$x_t(\rho\,\omega) = X_{\sigma^*(D_{k_0}) - t - 0}$$

with k_0 as above. Clearly ρ is bijective and Borel measurable and $\rho = \rho^{-1}$. Our general result on time reversal is

Theorem 5.3. For $\xi \geq 0$ on Ω

$$\mathcal{E}\,\xi \circ \rho = \mathcal{E}\,\xi . \quad ///$$

Proof. It suffices to consider

$$\xi = f_0(X_{\sigma(D_p)}) \cdots f_n(X_{\sigma(D_p) + t_n})$$

with $0 < t_1 < \cdots < t_n$ and with $f_i \geq 0$ and in $C_{com}(\underline{X})$. Then for k sufficiently large

$$\begin{aligned} \mathcal{E}\,\xi \circ \rho &= \mathcal{E}\,\xi \circ \rho_k \circ J_{\infty,k} \\ &= \mathcal{E}^{(k)}\,\xi \circ \rho_k \end{aligned}$$

which by Lemma 5.1

$$= \mathcal{S}^{(k)} \xi \circ \tau_k$$

$$= \mathcal{S} \xi \circ \tau_k \circ J_{\infty, k}$$

$$= \mathcal{S} \xi . \quad ///$$

From the very definition of $\mathcal{S}$

$$(5.8) \qquad \mathcal{S} \int_{\zeta^*}^{\zeta} dt \, \varphi(X_t) = \int dx \, \varphi(x)$$

for $\varphi \geq 0$ on $\underline{X}$. This identity will play an important role in later sections.

With the help of the second remark following Theorem 4.14 it is easy to show that a Borel subset E of X has finite capacity if and only if

$$H^E 1 = NL_E$$

with L_E a measure in $\mathcal{M}$ concentrated on the closure $c\ell(E)$ and then

$$\mathrm{Cap}(E) = \int L_E(dx).$$

For each k

$$\mathcal{P}^{(k)} [\sigma(E) < +\infty].$$

$$= \int L_k(dx) \, NL_E(x)$$

$$= \int L_E(dx) \, H^k 1(x)$$

and after passage to the limit $k \uparrow \infty$

$$(5.9) \qquad \mathrm{Cap}(E) = \mathcal{P}[\sigma(E) < +\infty].$$

This identity is also valid when $\mathrm{Cap}(E) = +\infty$.

6. Additive Functionals

Meyer's decomposition theory for supermartingales and for square integrable martingales plays an important role in this section. We cite [35] as a general reference.

We continue to work with a transient regular Dirichlet space $(\underline{F}, E)$.

For t a constant time on Ω or Ω_∞ the past $\mathcal{F}_t$ is the σ-algebra generated by the trajectory variables X_s, $s \leq t$. For τ a random time on Ω or Ω_∞ the past $\mathcal{F}_\tau$ is the σ-algebra of Borel subsets A of Ω such that

$$A \cap [\tau \leq t] \text{ is in } \mathcal{F}_t$$

for all constant times t. This terminology is consistent with standard usage.

6.1. Definition. An additive functional on the extended sample space Ω_∞ is a real valued function $\alpha(t, \omega)$ defined and jointly measurable for t real and for ω in Ω_∞ and satisfying the following conditions.

6.1.1. $\alpha(t, \cdot)$ is $\mathcal{F}_t$ measurable

6.1.2. $\alpha(t, \omega) = 0$ for $t < \zeta^*$

6.1.3. $\alpha(t, \omega) = \alpha(\zeta, \omega)$ for $t \geq \zeta$

6.1.4. $\alpha(t+h, \omega) - \alpha(t, \omega) = \alpha(t'+h, \omega') - \alpha(t', \omega')$

whenever $X_{t+s}(\omega) = X_{t'+s}(\omega')$ for $0 \leq s \leq h$. ///

6.2. Definition. An additive functional on the standard sample space Ω is a real valued function $\alpha(t, \omega)$ defined and jointly measurable for $t \geq 0$ and for ω in Ω and satisfying the following conditions.

6.2.1. $\alpha(t, \cdot)$ is $\mathcal{F}_t$ measurable.

6.2.2. $\alpha(t, \omega) = \alpha(\zeta, \omega)$ for $t \geq \zeta$.

6.2.3. $\alpha(t+h,\omega) - \alpha(t,\omega) = \alpha(t'+h,\omega') - \alpha(t',\omega')$ whenever $X_{t+s}(\omega) = X_{t'+s}(\omega')$ for $0 \leq s \leq h$. ///

Properties 6.1.4 and 6.2.3 guarantee that an additive functional is always "perfect" in the sense of [2]. In the constructions given below this can always be obtained by first selecting a sequence of approximating functionals for which the appropriate limits exist almost everywhere and then defining the limiting functional by an explicit limiting procedure involving this sequence which makes sense for all sample paths. We will take this for granted throughout the section.

We begin by constructing functionals on the extended sample space Ω_∞.

For $\varphi \geq 0$ on $\underline{\underline{X}}$ define

$$a(\varphi \cdot dx\,; t) = \int_{\zeta^*}^{t} ds\, \varphi(X_s)$$

on Ω_∞ with the understanding that $a(\varphi \cdot dx\,;\, t) = 0$ for $t < \zeta^*$. It is easy to check that

$$(6.1) \qquad \mathcal{E}\, a(\varphi \cdot dx;\, \zeta) = \int dx\, \varphi(x)$$

$$(6.2) \qquad \frac{1}{2}\, \mathcal{E}\, \{a(\varphi \cdot dx;\zeta)\}^2 = \int dx\, \varphi(x)\, G\varphi(x).$$

For μ in $\mathfrak{M}$ the functions $u\,G_u N\mu \to N\mu$ in $\underline{\underline{F}}_{(e)}$ and also increase to $N\mu$ quasi-everywhere as $u \uparrow \infty$. For typographical convenience put

$$\varphi_u = uN_u\mu.$$

Then for $0 < u < v$

$$\frac{1}{2}\, \mathcal{E}\, \{a(\varphi_v \cdot dx;\zeta) - a(\varphi_u \cdot dx\,;\zeta)\}^2$$

$$= \int dx\, \{\varphi_v(x) - \varphi_u(x)\}\, \{G\varphi_v(x) - G\varphi_u(x)\}$$

$$= E(G\varphi_v - G\varphi_u,\, G\varphi_v - G\varphi_u)$$

$$= E(vG_v N\mu - uG_u N\mu,\; v\,G_v N\mu - uG_u N\mu)$$

which $\to 0$ as $u, v \uparrow \infty$. Thus as $u \uparrow \infty$

$$a(\mu ; \zeta) = \text{Lim } a(\varphi_u \cdot dx; \zeta)$$

exists in the L^2 sense relative to $\mathcal{P}$ and therefore relative to $\mathcal{P}^{(k)} = \int L_k(dx) \mathcal{P}_x$.

For $t \geq 0$ clearly

$$\mathcal{E}^{(k)}(a(\varphi_u \cdot dx ; \zeta) \mid \mathcal{F}_t) = a(\varphi_u \cdot dx ; t) + G\varphi_u(X_t) .$$

By the maximal inequality for martingales

$$\sup_{t \geq 0} | \mathcal{E}^{(k)}(a(\varphi_v \cdot dx ; \zeta) | \mathcal{F}_t) - \mathcal{E}^{(k)}(a(\varphi_u \cdot dx; \zeta) | \mathcal{F}_t) | \to 0$$

in measure as $u, v \uparrow \infty$. (This supremum is taken either for $t \geq 0$ rational or for $t \geq 0$ real for a right continuous version.) The relevant estimate is independent of k and therefore

$$\sup_t | a(\varphi_v \cdot dx ; t) + G\varphi_v(X_t) - a(\varphi_u \cdot dx; t) - G\varphi_u(X_t) | \to 0$$

in measure relative to $\mathcal{P}$. (This supremum is over all real t.) It follows from (3.9) and (5.9) that also

$$\sup_t | G\varphi_v(X_t) - G\varphi_u(X_t) | \to 0$$

in measure relative to $\mathcal{P}$ as $u, v \uparrow \infty$. Thus after taking into account Meyer's uniqueness results for the decomposition of supermartingales

Theorem 6.1. For μ in $\mathfrak{M}$ there is a nonnegative additive functional $a(\mu ; t)$ on Ω_∞ which is unique up to $\mathcal{P}$ equivalence and satisfies the following conditions.

(i) Except for a $\mathcal{P}$ null set of sample paths, $a(\mu ; \cdot)$ is continuous and nondecreasing.

(ii) For each k and for $t \geq 0$

(6.3) $\mathscr{E}(a(\mu;\zeta) \mid \mathcal{F}_{\sigma(D_k)+t},\ \sigma(D_k) < +\infty)$

$$= a(\mu;\sigma(D_k)+t) + N\mu(X_{\sigma(D_k)+t}) . ///$$

The identities

(6.1′) $$\mathscr{E}\, a(\mu;\zeta) = \int \mu(dx)$$

(6.2′) $$\tfrac{1}{2}\,\mathscr{E}\,\{a(\mu;\zeta)\}^2 = E(\mu)$$

can be established either by applying (6.3) or by passing to the limit in (6.1) and (6.2).

6.3. Extension. Because of Lemma 3.18, Theorem 6.1 extends in an obvious way to general Radon measures μ which charge no polar set. We take this for granted below. ///

Consider again μ and φ_u as above. For ψ in $C_{com}(\underline{X})$ clearly $\psi(x)\,\varphi_u(x)\,dx \to \psi(x)\,\mu(dx)$ vaguely and the energies $E(\psi\varphi_u \cdot dx)$ are uniformly bounded. Thus $G(\psi\varphi_u) \to N(\psi\cdot\mu)$ weakly in $\underline{F}_{(e)}$. After possibly replacing the φ_u by Cesàro sums of a sequence (see the paragraph following 1.6.1′) we can assume that actually $G(\psi\varphi_u) \to N(\psi\cdot\mu)$ strongly in $\underline{F}_{(e)}$ and then the proof of Theorem 6.1 shows that

(6.4) $$\sup_t \left| a(\psi\varphi_u\cdot dx\,;\,t) - a(\psi\mu\,;t)\right| \to 0$$

in measure relative to $\mathscr{P}$. For a fixed trajectory ω and for a fixed real t the function $\psi(X_s)$, $\zeta^* \le s \le t$ can be uniformly approximated by a step function in s and from (6.4) without ψ it follows that

$$\int_{\zeta^*}^{t} ds\, \psi(X_s)\,\varphi_u(X_s) \to \int_{\zeta^*}^{t} a(\mu;\,ds)\,\psi(X_s)$$

and therefore

$$(6.5) \qquad a(\psi\cdot\mu;t) = \int_{\zeta^*}^{t} a(\mu;\, ds)\, \psi(X_s)$$

for almost every trajectory ω in Ω_∞ and for all real t. The collection of ψ for which this is valid is closed under monotonically increasing and bounded monotonically decreasing limits and there follows

Theorem 6.2. *Let μ be in $\mathfrak{M}$ and let $\psi \geq 0$ be a Borel function such that also $\psi\cdot\mu$ is in $\mathfrak{M}$. Then except for a $\mathcal{P}$ null set of trajectories ω in Ω_∞, the relation (6.5) is valid for all t.* ///

Remark. The proof of the corresponding Theorem 3.3 in [44] is incorrect. ///

Theorem 6.2 leads to a simple but important *property of universality* for the measure $\mathcal{P}$.

Theorem 6.3. *Let $\xi \geq 0$ on Ω and let μ be in $\mathfrak{M}$. Then*

$$(6.6) \qquad \int \mu(dx)\, \mathcal{E}_x \xi = \mathcal{E} \int_{\zeta^*}^{\zeta} a(\mu;\, dt)\, \xi\cdot\theta_t \,. ///$$

Of course the *shift* θ_t is interpreted as a mapping from Ω_∞ to Ω and is defined by

$$\theta_t\, \omega(s) = \omega(t+s).$$

To prove Theorem 6.3 it suffices to observe that if $\psi(x) = \mathcal{E}_x \xi$ then by the simple Markov property for the approximating $\mathcal{P}^{(k)}$ the right side of (6.6)

$$= \mathcal{E} \int_{\zeta^*}^{\zeta} a(\mu;\, dt)\, \psi(X_t)$$

$$= \mathcal{E}\, a(\psi\cdot\mu;\, \zeta)$$

$$= \int \mu(dx)\, \psi(x). ///$$

6.6

We turn now to another class of functionals which will play an important role in later sections. For $f = G\varphi$ with φ bounded and integrable and with f bounded (and therefore in $\underline{\underline{F}}_{(e)}$) we define on Ω_∞

$$(6.7) \qquad Mf(t) = I(\zeta^* \leq t < \zeta)\, f(X_t) + I(t > \zeta^*) \int_{\zeta^*}^{t} dx\, \varphi(X_s).$$

Clearly

$$(6.8) \qquad Mf(\zeta) = \int_{\zeta^*}^{\zeta} dx\, \varphi(X_s)$$

$$(6.9) \qquad Mf(\zeta^*) = I(\zeta^* > -\infty;\ X_{\zeta^*} \neq \partial)\, f(X_{\zeta^*})$$

$$(6.10) \qquad \tfrac{1}{2}\, \mathscr{E}\, \{Mf(\zeta)\}^2 = E(f, f).$$

Conditioned on the set $[\sigma(D_k) < +\infty]$ the process $\{Mf(\sigma(D_k)+t), t \geq 0\}$ is a martingale relative to the σ-algebras $\mathcal{F}_{\sigma(D_k)+t}$. In particular

$$I(\sigma(D_k) < +\infty)\, Mf(\sigma(D_k))\ ;\ I(\sigma(D_k) < +\infty)\, \{Mf(\zeta) - Mf(\sigma(D_k))\}$$

are mutually orthogonal and therefore

$$(6.11) \qquad \mathscr{E}[\sigma(D_k) < +\infty;\ \{Mf(\sigma(D_k))\}^2]$$

$$+\ \mathscr{E}[\sigma(D_k) < +\infty;\ \{Mf(\zeta) - Mf(\sigma(D_k))\}^2]$$

$$= \mathscr{E}[\sigma(D_k) < +\infty;\ \{Mf(\zeta)\}^2]$$

$$\leq 2\, E(f, f).$$

As $k \uparrow \infty$ clearly

$$(6.12) \qquad I[\sigma(D_k) < +\infty]\, Mf(\sigma(D_k)) \to Mf(\zeta^*)$$

and it follows that

$$Mf(\zeta^*)\ ;\ Mf(\zeta) - Mf(\zeta^*)$$

are square integrable and mutually orthogonal and therefore

$$(6.13)\qquad \frac{1}{2}\mathcal{E}\{Mf(\zeta^*)\}^2 + \frac{1}{2}\mathcal{E}\{Mf(\zeta) - Mf(\zeta^*)\}^2$$

$$= \frac{1}{2}\mathcal{E}\{Mf(\zeta)\}^2$$

$$= E(f,f) .$$

Also the convergence in (6.12) must be in mean square. With the help of the maximal inequality for square integrable martingales [35, p.88] it is easy now to pass to the limit in f and establish

Theorem 6.4. *For f in $\underline{F}_{(e)}$ there is an additive functional Mf(t) on Ω_∞ which is unique up to $\mathcal{P}$ equivalence and satisfies the following conditions.*

(i) The difference $Mf(t) - f(X_t)$ is continuous except for a $\mathcal{P}$ null set of sample paths.

(ii) Conditioned on the set $[\sigma(D_k) < +\infty]$ the process $\{Mf(\sigma(D_k)+t), t \geq 0\}$ is a martingale relative to the σ-algebras $\mathcal{F}_{\sigma(D_k)}$.

(iii) Mf is given by (6.7) when $f = G\varphi$ and (6.13) is valid for general f in $\underline{F}_{(e)}$ with convergence in (6.12) both in the [a.e. $\mathcal{P}$] and $L^2(d\mathcal{P})$ sense. ///

Before continuing we note that for the special case $f = G\varphi$ we have

$$f(X_t) = Mf(\zeta^*) + Mf(t) - Mf(\zeta^*) - \int_{\zeta^*}^{t} ds\, \varphi(X_s)$$

and therefore

$$\sup_t |f(X_t)| \leq |Mf(\zeta^*)| + \int_{\zeta^*}^{\zeta} dx\, |\varphi(X_s)|$$

$$+ \sup_t |Mf(t) - Mf(\zeta^*)| .$$

This together the maximal inequality for martingales gives the crude but useful estimate.

$$(6.14) \qquad \mathscr{E}\ \sup_t |f(X_t)|^2 \leq 24\, E(f,f) + 3\,\mathscr{E}\, \left| \int_{\zeta^*}^{\zeta} ds\, |\varphi(X_s)| \right|$$

Also we note the easily verified relation

$$(6.15) \qquad Mf(t) = f(X_t) + a(\mu\,;\ t)$$

for the special case $f = N\mu$ with μ in $\mathfrak{M}$.

The next theorem is an immediate consequence of Theorem 6.4-(ii) and of Meyer's result on increasing processes associated with a square integrable martingale.

Theorem 6.5. For each f in $\underline{F}_{(e)}$ there is an additive functional $\langle Mf\rangle(t)$ on Ω_∞ which is unique up to $\mathscr{P}$ equivalence and satisfies the following conditions

(i) $\langle Mf\rangle(t)$ is continuous and nondecreasing for $t \geq \zeta^*$. Also $\langle Mf\rangle(\zeta^*) = I(X_{\zeta^*} \neq \partial)\, f^2(X_{\zeta^*})$.

(ii) Conditioned on the set $[\sigma(D_k) < +\infty]$ the process $\{ (Mf(\sigma(D_k)+t))^2 - \langle Mf\rangle(\sigma(D_k)+t),\ t \geq 0 \}$ is a martingale. ///

We introduce also $M_c f(t)$, the continuous part of $Mf(t)$ and the nondecreasing functional $\langle M_c f\rangle(t)$ having the same relation to $M_c f(t)$ that $\langle Mf\rangle(t)$ has to $Mf(t)$. It follows from (6.13) and from Meyer's results [35, VIII.3] that

$$(6.16) \qquad E(f,f) = \frac{1}{2}\,\mathscr{E}\langle M_c f\rangle(\zeta) + \frac{1}{2}\,\mathscr{E}\,\Sigma_t\, \{f(X_t) - f(X_{t-0})\}^2 .$$

Note that the sum on the right includes the terms $I(\zeta^* > -\infty; X_{\zeta^*} \neq \partial) f^2(X_{\zeta^*})$ and $I(\zeta < +\infty;\ X_{\zeta-0} \neq \partial)\ f^2(X_{\zeta-0})$.

The functionals $\langle Mf \rangle$ and $\langle M_c f \rangle$ will be referred to below as <u>Dirichlet functionals.</u> Both will be used. The functional $\langle Mf \rangle$ is convenient for calculations but $\langle M_c f \rangle$ is often better for stating results since its increments are invariant under time reversal.

Finally we note that the above functionals are also well defined on the standard sample space Ω. The main tool for showing this is the property of universality (6.6). To see how this works define

$$(6.17) \qquad Mf(t) = f(X_t) + \int_0^t ds\, \varphi(X_s)$$

when $f = G\varphi$ in $\underline{\underline{F}}_{(e)}$. For general f in $\underline{\underline{F}}_{(e)}$ choose φ_n such that $G\varphi_n \to f$ in $\underline{\underline{F}}_{(e)}$ and such that except for a $\mathcal{P}$ null subset of Ω_∞ the functionals $MG\varphi_n(t)$ converge uniformly in t as $n \uparrow \infty$. If ξ is the indicator of the set where $MG\varphi_n$ does not converge uniformly then by (6.6)

$$\int \mu(dx)\, \mathcal{E}_x\, \xi = 0$$

for all μ in $\mathcal{M}$ and it follows from Corollary 3.12 that for quasi-every x also $MG\varphi_n(t)$ converges uniformly except for a $\mathcal{P}_x$ null subset of Ω. Another application of (6.6) shows that for quasi-every x the limiting process is a square integrable martingale. Similar arguments work for the other functionals introduced above. From now on we take for granted the definition of these functionals on Ω. We note in particular the relation

$$(6.18) \qquad N_u \nu(x) = \mathcal{E}_x \int_0^\zeta a(\nu;\, dt)\, e^{-ut} \qquad u \geq 0.$$

7. Balayage

We continue to assume that $(\underline{F}, E)$ is transient.

7.1. Definition. Let M be a Borel subset of $\underline{X}$. A point x is regular for M if $\mathcal{P}_x[\sigma^+(M) = 0] = 1$. The set M^r is the subset of x in $\underline{X}$ which are regular for M. ///

We have already noted (see Remark 1 following Theorem 4.14) that the difference $M - M^r$ is polar. Also it follows from Theorem 4.14 that M^r is Borel measurable modulo a polar set.

7.2. Definition. A Borel subset M of $\underline{X}$ is finely closed if the difference $M^r - M$ is polar.///

Of course every closed set is finely closed. In general there exist finely closed sets which are not closed, even modulo a polar set

We begin with

Lemma 7.1. Let M be a finely closed Borel subset of $\underline{X}$ and let $\mathcal{M}(M)$ be the collection of measures μ in $\mathcal{M}$ which do not charge $\underline{X}-M$. Then $\mathcal{M}(M)$ is convex and closed relative to the energy metric E. ///

Proof. Convexity is obvious. To show that $\mathcal{M}(M)$ is closed observe that μ in $\mathcal{M}$ belongs to $\mathcal{M}(M)$ if and only if $\int \mu(dx) \exp\{-u\,\sigma(M)\}$ is independent of $u > 0$. Since μ can be replaced by $\psi \cdot \mu$ with $\psi \geq 0$ in $C_{com}(\underline{X})$, we can assume that M has compact closure. But then the integrand belongs to $\underline{F}_{(e)}$ (see Remark 2 following Theorem 4.14) and therefore the above condition is preserved by convergence in $\mathcal{M}$. ///

7.3. Notation. For M a finely closed Borel set let $\mathcal{M}(M)$ be as in Lemma 14.1 and let $[N\mathcal{M}(M)]$ be the closed linear subspace spanned by $N\mathcal{M}(M)$. Standard Hilbert space arguments establish for each μ in M the existence of a unique measure $\Pi^M\mu$, the balayage of μ onto M, such that $\Pi^M\mu$ belongs to $\mathcal{M}(M)$ and such that $E(\mu - \Pi^M\mu)$ is minimal. ///

The balayaged measure $\Pi^M\mu$ is characterized in the following proposition. The proof is essentially that of Cartan [4].

Proposition 7.2. Let M be finely closed and let μ be in $\mathcal{M}$.

(i) $N\,\Pi^M\mu$ is the E orthogonal projection of $N\mu$ onto $[N\mathcal{M}(M)]$.

(ii) $N\mu \geq N\Pi^M\mu$ quasi-everywhere and $N\mu = N\Pi^M\mu$ quasi-everywhere on M.

(iii) $N\Pi^M\mu$ is the unique element with minimal E norm among f in $\underline{F}_{(e)}$ such that $f \geq N\mu$ quasi-everywhere on M. ///

Proof. From the relation $E(\mu - \Pi^M\mu) \leq E(\mu-\lambda)$ for all λ in $\mathcal{M}(M)$ follows

$$E(N\mu-N\Pi^M\mu,\ N\lambda - N\Pi^M\mu) \leq 0$$

for λ in $\mathcal{M}(M)$. Taking $\lambda = 0$ and $\lambda = 2\,\Pi^M\mu$ we deduce that

$$E(N\mu - N\Pi^M\mu,\ N\Pi^M\mu) = 0 \tag{7.1}$$

and therefore

$$E(N\mu - N\Pi^M\mu,\ N\lambda\,) \leq 0 \tag{7.2}$$

for λ in $M(\nu)$. The inequality (7.2) implies that $N\Pi^M\mu \geq N\mu$ quasi-everywhere on M. Then (7.1) implies that $N\Pi^M\mu = N\mu$ [a.e. $\Pi^M\mu$] and therefore by the maximum principle (Corollary 3.14) $N\Pi^M\mu \leq N\mu$ quasi-everywhere and (ii) is proved. Finally (i) follows from (ii) and (3.10) and

(iii) follows since

$$E(f,f) = E(N\Pi^M\mu, N\Pi^M\mu) + E(f - N\Pi^M\mu, f - N\Pi^M\mu)$$

$$+ 2\int \Pi^M\mu(dx)\{f(x) - N\mu(x)\}. ///$$

For M Borel and finely closed denote the <u>complement</u> X-M by D and consider the corresponding <u>absorbed process</u>

$$X_t^D = \begin{cases} X_t & \text{for } t < \sigma(M) \\ \partial & \text{for } t \geq \sigma(M). \end{cases}$$

We are interested primarily in the corresponding resolvent operators

$$G_u^D f(x) = \mathcal{E}_x \int_0^{\sigma(M)} dt\, e^{-ut} f(X_t) \qquad u \geq 0$$

and their connection with the hitting operators H_u^M. In particular we note the familiar and easily established identity

$$(7.3) \qquad G_u = G_u^D + H_u^M G_u \qquad u \geq 0.$$

The basic results are collected in

<u>Theorem 7.3.</u> <u>Let M be a finely closed Borel set and let</u> $D = \underline{X} - M$.

(i) <u>For</u> f <u>in</u> $\underline{F}_{(e)}$ <u>the function</u> $H^M f$ <u>is the quasi-continuous version of the E orthogonal projection of f onto the linear subspace</u> $[N\mathfrak{M}(M)]$.

(ii) <u>For</u> μ <u>in</u> M <u>and for</u> $\varphi \geq 0$ <u>on</u> M

$$(7.4) \qquad \int \Pi^M\mu(dy)\,\varphi(y) = \int \mu(dx)\, H^M\varphi(x).$$

(iii) <u>The operators</u> G_u^D, $u > 0$ <u>form a symmetric resolvent on</u> $L^2(D, dx)$. <u>The corresponding Dirichlet space</u> $(\underline{F}^D, E^D)$ <u>is given by</u>

$$\underline{\underline{F}}^D = \{f \text{ in } F : f = 0 \quad \text{q.e. on } M\}.$$

$$E^D(f,g) = E(f,g) \qquad\qquad f, g \text{ in } \underline{\underline{F}}^D$$

<u>This absorbed Dirichlet space</u> $(\underline{\underline{F}}^D, E^D)$ <u>is transient and the associated extended Dirichlet space is</u>

$$\underline{\underline{F}}^D_{(e)} = \{f \text{ in } \underline{\underline{F}}_{(e)} : \; f = 0 \quad \text{q.e. on } M\} \quad .$$

<u>Moreover</u> $\underline{\underline{F}}^D_{(e)}$ <u>is the E orthogonal complement of</u> $[N\mathcal{M}(M)]$ <u>in</u> $F_{(e)}$.

(iv) <u>Let</u> μ <u>in</u> $\mathcal{M}$ <u>be concentrated on D and equivalent to dx on D</u>. <u>If</u> $\Gamma \subset M$ <u>is null for</u> $\Pi^M\mu$ <u>then</u> $H^M(x,\Gamma) = 0$ <u>for quasi-every x in</u> D. <u>Also</u> $\Pi^M\nu$ <u>is absolutely continuous relative to</u> $\Pi^M\mu$ <u>whenever</u> ν <u>in</u> $\mathcal{M}$ <u>is concentrated on</u> D. ///

<u>Proof.</u> We begin with (i). It suffices to consider the special case of a bounded potential $f = N\mu$. The estimates

$$(7.5) \qquad\qquad H^M f \le f$$

$$(7.6) \qquad\qquad P_t H^M f \le H^M f$$

follow from the supermartingale property of $\{f(X_t), t \ge 0\}$. At this point we cannot conclude directly from (7.5) and (7.6) that $H^M f$ belongs to $\underline{\underline{F}}_{(e)}$. (See the remark following Lemma 3.3.) Therefore we must proceed indirectly, <u>considering first the special case</u> when $\underline{\underline{F}}_{(e)} = \underline{\underline{F}}$ so that we can conclude from Lemma 3.3 that $H^M f$ is again a potential.

Since $H^M f = f = N\Pi^M\mu$ quasi-everywhere on M the difference $H^M f - N\Pi^M\mu$ is E orthogonal to $[N\mathfrak{M}(M)]$ and therefore

$$(7.7) \qquad E(H^M f, H^M f) = E(N\Pi^M\mu, N\Pi^M\mu) + E(H^M f - N\Pi^M\mu, H^M f - N\Pi^M\mu).$$

Also $H^M f = H^M N\Pi^M\mu$ satisfies $H^M f \leq N\Pi^M\mu$ by (7.5) with $N\Pi^M\mu$ playing the role of f and so again by Lemma 3.3

$$(7.8) \qquad E(H^M f, H^M f) \leq E(N\Pi^M\mu, N\Pi^M\mu)$$

and (i) for this special case follows from (7.7), (7.8) and Proposition 7.2-(i). To strip away the hypothesis $\underline{\underline{F}}_{(e)} = \underline{\underline{F}}$ it suffices to show that $H^M f$ belongs to $\underline{\underline{F}}_{(e)}$ in the general case. For this purpose choose approximating f_n as in Definition 1.6 which are uniformly bounded. For fixed n and for $u > 0$ the special case applies and so

$$E_u(H_u^M f_n, H_u^M f_n) \leq E_u(f_n, f_n).$$

Thus for $0 < u < v$

$$E(H_u^M f_n, H_u^M f_n) \leq E_v(f_n, f_n)$$

and after passing to the limit $u \downarrow 0$ and then $v \downarrow 0$

$$(7.9) \qquad E(H^M f_n, H^M f_n) \leq E(f_n, f_n) .$$

We can assume that $f_n \to f$ quasi-everywhere and then it follows from Proposition 4.15 that $f_n \to f$ [a.e. $H^M(x, d\cdot)$] for quasi-every x and therefore $H^M f_n \to H^M f$ quasi-everywhere. This together with the estimate (7.9) guarantees that $H^M f$ is in $\underline{\underline{F}}_{(e)}$ and (i) is completely proved. Conclusion (ii) follows from (i) since if φ is in $\underline{\underline{F}}_{(e)}$ then the left side of (7.4) $= E(N\Pi^M\mu, \varphi) = E(H^M N\mu, \varphi) = E(N\mu, H^M\varphi) =$ the right side of (7.4). We turn next to (iii). The resolvent identities can be established by a straightforward computation which we omit. Symmetry follows from symmetry for the resolvent operators G_u, from (7.3) and from symmetry of $H^M_u G_u$ as an operator on $L^2(dx)$ which is an immediate consequence of the $u > 0$ versions of conclusion (i). The remainder of (iii) follows directly from (i) and from (7.3) for $u = 0$. To prove (iv) consider Γ a $\Gamma^M\mu$ null subset of M. Then by (7.4) we have $H^M(x,\Gamma) = 0$ almost everywhere on D. But then by Lemma 4.6 $uG^D_u H^M(\cdot,\Gamma) = 0$ quasi-everywhere and since $uG^D_u H^M(\cdot,\Gamma) \uparrow H^M(\cdot,\Gamma)$ quasi-everywhere on D as $u \uparrow \infty$ also $H^M(\cdot,\Gamma) = 0$ quasi-everywhere on D. The remainder of (iv) follows with the help of (7.4). ///

The $u > 0$ versions of results in this section are valid also in the recurrent case. We take this for granted throughout the volume. Also we note for future use the resolvent like identity

$$(7.10) \qquad H^M_u = H^M_v + (v-u)\, H^M_u G^D_v$$

valid for $u, v \geq 0$, which is easily established by direct computation.

8. Random Time Change

In this section $(\underline{F}, E)$ is either transient or recurrent.

Fix a nontrivial Radon measure ν charging no polar set. Then for quasi-every x

(8.1) $$\mathcal{P}_x[a(\nu;t) = +\infty \text{ for some } t < \zeta] = 0.$$

Consider the inverse process

(8.2) $$b(\nu;s) = \inf\{t > 0 : a(\nu;t) > s\}$$

with the understanding that $b(\nu;s) = +\infty$ when not otherwise defined. Let M be the set of x such that

$$\mathcal{P}_x[b(\nu;0) = 0] = 1.$$

It is easy to see that M is finely closed and that

(8.3) $$\mathcal{P}_x[b(\nu;0) = \sigma(M)] = 1$$

for quasi-every x. We study the time changed process

$$X_t^\nu = X_{b(\nu;t)} .$$

We are interested not so much in the process as in the time changed resolvent operators

$$R_\alpha^\nu \varphi(x) = \mathcal{E}_x \int_0^\infty dt\, e^{-\alpha t} \varphi(X_t^\nu)$$

and the associated Dirichlet space $(\underline{H}, Q)$. An elementary change of variables along sample paths establishes the formula

$$R_\alpha^\nu \varphi(x) = \mathcal{E}_x \int_0^\infty a(\nu;\, dt) e^{-\alpha a(\nu;t)} \varphi(X_t).$$

We will also be interested in the modified resolvent operators

$$R^{\nu}_{(u)\alpha}\varphi(x) = \mathcal{E}_x \int_0^\infty a(\nu;dt)\, e^{-ut - \alpha a(\nu;t)} \varphi(X_t). \qquad 8.2$$

We begin with

<u>Theorem 8.1.</u> <u>The family</u> $\{R^{\nu}_{\alpha}, \alpha > 0\}$ <u>and also for</u> $u > 0$ <u>the family</u> $\{R^{\nu}_{(u)\alpha}, \alpha > 0\}$ <u>is a symmetric submarkovian resolvent on</u> $L^2(\nu)$. ///

<u>Proof.</u> It suffices to consider $R_{(u)\alpha}$ since the analogous results for R^{ν}_{α} can be established by passage to the limit $u \downarrow 0$. A simple computation gives $\alpha R^{\nu}_{(u)\alpha} 1 \leq 1$. To establish symmetry let $\mathcal{E}_{(u)}$ be the approximate Markov process of Section 5 which corresponds to $(\underline{F}, E_u)$. Then for $\varphi, \psi \geq 0$ on M

$$\int \nu(dx)\, \psi(x)\, R^{\nu}_{(u)\alpha}\varphi(x)$$

$$= \int \nu(dx)\, \psi(x)\, \mathcal{E}_x \int_0^\zeta a(\nu;\, dt) e^{-ut - \alpha a(\nu;\, t)}\, \varphi(X_t)$$

which by (6.6)

$$= \mathcal{E}_{(u)} \int_{\zeta *}^{\zeta} a(\nu;\, ds)\, \psi(X_s) \int_s^\zeta a(\nu; dt)\, \varphi(X_t) e^{-\alpha a(\nu;t) + \alpha a(\nu;s)}$$

and symmetry follows from the invariance of $\mathcal{E}_{(u)}$ under time reversal, Theorem 5.3. Finally the resolvent identity follows easily from

$$\int_0^{R_u} a(\nu;\, dt)\, \{ e^{-\alpha a(\nu;\, t)} - e^{-\beta a(\nu;\, t)} \}\, \varphi(X_t)$$

$$= \int_0^{R_u} a(\nu;\, dt) \varphi(X_t) e^{-\alpha a(\nu;\, t)} \{1 - e^{-(\beta - \alpha) a(\nu;\, t)}\}$$

$$= \int_0^{R_u} a(\nu; dt) \varphi(X_t) e^{-\alpha a(\nu;\, t)} (\beta - \alpha) \int_0^t a(\nu;\, ds) e^{-(\beta - \alpha)\, a(\nu;s)}$$

$$= (\beta - \alpha) \int_0^{R_u} a(\nu; s) e^{-\alpha a(\nu;s)} \int_s^{R_u} a(\nu; dt) \varphi(X_t) e^{-\alpha[a(\nu;\, t) - a(\nu;\, s)]}$$

where R_u is the usual terminal variable exponentially distributed at the rate u. ///

Note that in the recurrent case the approximate Markov process $\mathcal{S}_{(u)}$ can be defined directly by

$$\mathcal{S}_{(u)} \xi = u \int dx \ \mathcal{S}_x^{(u)} \ \xi$$

where $\mathcal{S}_x^{(u)}$ is obtained by "killing the trajectory" at the random time R_u.

8.1. Notation. Restriction to M is denoted by γ. ///

We turn now to the problem of identifying the time changed Dirichlet space. The transient case is easily disposed of.

Theorem 8.2. If $(\underline{F}, E)$ is irreducible and transient then also the time changed Dirichlet space $(\underline{H}, Q)$ is irreducible and transient. Moreover $\underline{H}_{(e)} = \gamma \underline{F}_{(e)}$ and

(8.4) $$Q(\varphi, \varphi) = E(H^M \varphi, \ H^M \varphi)$$

for φ in $\gamma \underline{F}_{(e)}$. ///

Proof. The key to the theorem is

(8.5) $$R^\nu \psi = \gamma N(\psi \cdot \nu)$$

which follows from (6.5) and (6.1'). Irreducibility is a consequence of Corollary 4.7-(ii). If $\nu(A) > 0$ then $N1_A \nu > 0$ quasi-everywhere and so $R^\nu 1_A > 0$ [a.e. ν]. Transience then follows from Lemma 3.18. Now consider ψ in $L^1(\nu)$ such that $N(\psi \cdot \nu)$ is bounded. Clearly $\varphi = \gamma N(\psi \cdot \nu) = R^\nu \psi$ is in $\gamma \underline{F}_{(e)}$ and in $\underline{H}_{(e)}$ and

$$Q(\varphi,\varphi) = E(N(\psi\cdot\nu)\ ,\ N(\psi\cdot\nu)\ ).$$

If we knew that $\nu(\underline{\underline{X}} - M) = 0$, then we could be sure that $N(\psi\cdot\nu)$ is in $[N\mathcal{M}(M)]$ and therefore $N(\psi\cdot\nu) = H^M\varphi$ and (8.4) would be true in this special case. Such functions are certainly dense in $\underline{\underline{H}}_{(e)}$ and so the theorem would be proved if we knew in addition that such functions are dense in $[N\mathcal{M}(M)]$. In checking that $\nu(\underline{\underline{X}} - M) = 0$ we can consider instead approximations from below by measures in $\mathcal{M}$ and so there is no loss of generality in assuming that ν is in $\mathcal{M}$. By (6.18) and (6.5) we have for quasi-every x

$$N_1(\nu\cdot\mathcal{E}_{\cdot}\, e^{-b(\nu;\,0)}\,)$$

$$= \mathcal{E}_x \int_0^\zeta a(\nu;dt)e^{-t}\, \mathcal{E}_{X_t}\, e^{-b(\nu;\,0)}$$

$$= \mathcal{E}_x \int_0^\zeta a(\nu;dt)e^{-t-\theta_t b(\nu;0)}$$

Along individual sample paths $t + \theta_t b\,(\nu;\,0) = t$ except for a t-set which is null for $a(\nu;\,\cdot)$ and so the last expression

$$= \mathcal{E}_x \int_0^\zeta a(\nu;dt)\ e^{-t}$$

$$= N_1\nu\,(x)$$

which implies that $\nu(\underline{\underline{X}} - M) = 0$. In checking that functions $N(\psi\cdot\nu)$ as above are dense in $[N\mathcal{M}(M)]$ it suffices to show that if f in $\underline{\underline{F}}_{(e)}$ is not orthogonal to $[N\mathcal{M}(M)]$ then it cannot be true that $f = 0$ [a.e. ν]. But the former guarantees that for some μ in $\mathcal{M}$, $\int \mu(dx)\, H^M|f|\,(x) > 0$ and so by right continuity of $f(X_t)$ and by (8.3) also $\int\mu(dx)\, N(|f|\cdot\nu)\,(x) =$ $\int\mu\,(dx)\,\mathcal{E}_x \int_0^\zeta a(\nu\,;dt)|f|\,(X_t) > 0$ and so $|f|\cdot\nu \neq 0$. ///

We consider now the case when $(\underline{F}, E)$ is recurrent. The result is the same but our proof is more involved . Recall that the extended Dirichlet space $\underline{F}_{(e)}$ is still well defined. However at this point general functions in $\underline{F}_{(e)}$ are specified and finite only up to dx equivalence. We will prove below that in fact all functions in $\underline{F}_{(e)}$ have quasi-continuous refinements.

Lemma 8.3. Let $(\underline{F}, E)$ be irreducible and recurrent.

(i) The time changed Dirichlet space $(\underline{H}, Q)$ is irreducible and recurrent.

(ii) The absorbed Dirichlet space $(\underline{F}^D, E)$ is transient.

(iii) If f is in $\underline{F}_{(c)}$ then $H^M f$ converges almost everywhere. Moveover $H^M f$ is in $\underline{F}_{(e)}$, the difference $f - H^M f$ is in $\underline{F}^D_{(e)}$ and

(8.6) $$E(f, f) = E(H^M f, H^M f) + E(f - H^M f, f - H^M f). \;///$$

Proof. Irreducibility follows as in the transient case since if $N_1 1_A \nu > 0$ then also $R_1^\nu 1_A > 0$. Recurrence follows from Corollary 4.7-(iii). This proves (i) and (ii) follows from Proposition 4.16 and Theorem 1.6 since for ν nontrivial the set M is nonpolar. (This application of Theorem 1.6 is valid even though $(\underline{F}^D, E)$ need not be irreducible.) In proving (iii) we will make constant use of (7.10) and Theorem 7.3. For f in $\underline{F}$ and for $0 < u < v$

(8.7) $$E_u(H_u^M f - H_v^M f, H_u^M f - H_v^M f)$$

$$= (v-u)\, E_u(G_u^D H_v^M f,\; H_u^M f - H_v^M f)$$

$$= (v-u) \int_D dx\, H_v^M f(x) \,\{ H_u^M f(x) - H_v^M f(x) \}$$

and in particular for $f \geq 0$

$$(8.8)\qquad E(H_u^M f - H_v^M f, H_u^M f - H_v^M f)$$

$$\leq v \int_D dx\, H_v^M f(x) \{ H^M f(x) - H_v^M f(x) \}.$$

Also

$$E_v(H_v^M f,\ H_v^M f)$$

$$= E_v(H_u^M f, H_v^M f)$$

$$= E_u(H_u^M f,\ H_v^M f) + (v-u) \int dx\, H_u^M f(x)\, H_v^M f(x)$$

and therefore

$$(8.9)\qquad E_v(H_v^M f, H_v^M f) - E_u(H_u^M f, H_u^M f)$$

$$= (v-u) \int dx\, H_u^M f(x)\, H_v^M f(x).$$

From this is follows in particular that

$$(8.10)\qquad \int dx\, H^M |f|\,(x)\, H_v^M |f|\,(x) < +\infty$$

and that

$$(8.11)\qquad \operatorname{Lim}_{v \downarrow 0} v \int dx\, H^M |f|\,(x)\, H_v^M |f|(x) = 0$$

for f in $\underline{F}$. But then the right side of (8.8) converges to 0 as $u, v \downarrow 0$ and we conclude that $\{H_u^M f,\ u > 0\}$ is Cauchy relative to E as $u \downarrow 0$. But then $\{f - H_u^M f, u > 0\}$ is Cauchy in the <u>transient</u> Dirichlet space $\underline{F}^D$. Thus $H^M f$ converges almost everywhere and belongs to $\underline{F}_{(e)}$. Also since

$$E_u(H_u^M f, H_u^M f) = E(H_u^M f, H_u^M f) + u \int dx \{H_u^M f(x)\}^2$$

it follows from (8.11) that

$$\text{(8.12)} \qquad E(H^M f, H^M f) = \lim_{u \downarrow 0} E_u(H_u^M f, H_u^M f).$$

Therefore

$$\begin{aligned} E(f,f) &= \lim_{u \downarrow 0} E_u(f,f) \\ &= \lim_{u \downarrow 0} \{E_u(H_u^M f, H_u^M f) + E_u(f - H_u^M f, f - H_u^M f)\} \\ &= \lim_{u \downarrow 0} E_u(H_u^M f, H_u^M f) + \lim_{u \downarrow 0} E(f - H_u^M f, f - H_u^M f) \\ &\quad + \lim_{u \downarrow 0} u \int dx \{f(x) - H_u^M f(x)\}^2 \\ &= E(H^M f, H^M f) + E(f - H^M f, f - H^M f) \end{aligned}$$

since

$$u \int dx \{f(x) - H_u^M f(x)\}^2 \le 2u \int dx\, f^2(x) + 2u \int dx \{H_u^M f(x)\}^2$$

converges to 0 as $u \downarrow 0$ and (iii) is proved for the special case of f in $\underline{F}$. Finally it is easy to extend (iii) to general f in $\underline{F}_{(e)}$ with the help of (8.6) and transience of $(\underline{F}^D, E)$. ///

Before continuing we note that a passage to the limit $u \downarrow 0$ in (8.9) gives

$$\text{(8.13)} \qquad E_v(H_v^M f, H_v^M f) = E(H^M f, H^M f) + v \int dx\, H^M f(x)\, H_v^M f(x)$$

for f in $\underline{F}$.

Applying Theorem 8.2 to $(\underline{F}, E_u)$ for $u > 0$ we get

Lemma 8.4. Let $(\underline{F}, E)$ be either transient or recurrent. For $u > 0$ the modified time changed Dirichlet space $(\underline{H}_{(u)}, Q_{(u)})$ which corresponds to $\{R^{\nu}_{(u)\alpha}, \alpha > 0\}$ is transient. Moreover $\underline{H}^{M}_{(u)(e)} = \gamma\,\underline{F}$ and

$$Q^{M}_{(u)}(\varphi, \varphi) = E_u(H^{M}_{u}\varphi, H^{M}_{u}\varphi)$$

for φ in $\gamma\,\underline{F}$. ///

Finally we are ready to pin down the time changed Dirichlet space in the recurrent case. Note first that since R^{M}_{α} dominates $R^{M}_{(u)\alpha}$, the function space $\underline{H}^{M}_{(e)}$ contains $\underline{H}^{M}_{(u)(e)}$ by Lemma 1.1 and so by Lemma 8.4 also $\underline{H}_{(e)}$ contains $\gamma\underline{F}$. Now fix $\alpha > 0$ and $\psi \geq 0$ both bounded and in $L^1(\nu)$ and note that $H^M R^{\nu}_{(u)\alpha}\psi \to H^M R^{\nu}_{\alpha}\psi$ quasi-everywhere as $u \downarrow 0$ and also

$$
\begin{aligned}
(8.14)\qquad & E(H^M R^{\nu}_{(u)\alpha}\psi,\ H^M R^{\nu}_{(u)\alpha}\psi) \\
&\leq E(H^{M}_{u} R^{\nu}_{(u)\alpha}\psi,\ H^{M}_{u} R^{\nu}_{(u)\alpha}\psi) \\
&\leq E_u(H^{M}_{u} R^{\nu}_{(u)\alpha}\psi,\ H^{M}_{u} R^{\nu}_{(u)\alpha}\psi) \\
&= Q_{(u)}(R^{\nu}_{(u)\alpha}\psi,\ R^{\nu}_{(u)\alpha}\psi) \\
&\leq Q_{(u)\alpha}(R^{\nu}_{(u)\alpha}\psi,\ R^{\nu}_{(u)\alpha}\psi) \\
&= \int \nu(dy)\, \psi(y)\, R^{\nu}_{(u)\alpha}\psi(y) \\
&\leq \int \nu(dy)\, \psi(y)\, R^{\nu}_{\alpha}\psi(y)
\end{aligned}
$$

which is bounded independent of $u > 0$. Thus $H^M R^\nu_\alpha \psi$ is in $\underline{\underline{F}}_{(e)}$ and therefore $R^\nu_\alpha \psi$ is in $\gamma\, \underline{\underline{F}}_{(e)}$. Next assume for the moment that D has finite measure. Then for θ in $\gamma\,(\underline{\underline{F}} \cap C_{com}(\underline{\underline{X}}))$ and therefore in $\underline{\underline{H}}_{(e)} \cap L^2(\nu) = \underline{\underline{H}}$

$$Q_\alpha(R_\alpha \psi, \theta)$$

$$= \int dx\ \psi(x)\theta(x)$$

$$= Q_{(u)}(R^\nu_{(u)\alpha}\psi, \theta)$$

$$= E_u(H^M_u R^\nu_{(u)\alpha}\psi,\ H^M_u \theta)$$

which by (8.13) is equal to

$$E(H^M R^\nu_{(u)\alpha}\psi, H^M \theta) \tag{8.15}$$

$$+ u \int dx\ H^M R^\nu_{(u)\alpha}\psi(x)\ H^M_u \theta(x).$$

Because of the estimate (8.14) we can assume, after possibly replacing the $H^M R^\nu_{(u)\alpha}\psi$ by Cesàro sums for a sequence of $u \downarrow 0$ (see the paragraph following 1.6.1') that the first term in (8.15) converges to $E(H^M R^\nu_\alpha \psi, H^M \theta)$ as $u \downarrow 0$. The second term is dominated by

$$u \int_D dx\ (1/\alpha)\ \|\psi\|_\infty \|\theta\|_\infty$$

$$+ u \int_M dx\ (1/\alpha)\ \|\psi\|_\infty \theta(x)$$

which converges to 0 because of our assumptions on θ and D and therefore

$$Q_\alpha(R^\nu_\alpha \psi, \theta) = E(H^M R^\nu_\alpha \psi, H^M \theta). \tag{8.16}$$

From (8.16) and our previous results that $\underline{\underline{H}}^M_{(e)}$ contains $\gamma\underline{\underline{F}}$ and that $R^\nu_a \psi$ is in $\gamma\underline{\underline{F}}_{(e)}$ it follows that $\underline{\underline{H}}_{(e)} = \gamma\underline{\underline{F}}_{(e)}$ and that (8.4) is valid for φ in $\gamma\underline{\underline{F}}_{(e)}$. To remove the restriction on D let $q(x)$ be integrable and everywhere > 0 and consider the time changed process $X^\sim_t$ which corresponds to $\nu^\sim = q \cdot dx$. For this time changed process D is empty and so by the above results the associated extended Dirichlet space is again $(\underline{\underline{F}}_{(e)}, E)$. But the original time changed process X^ν_t can also be obtained by time changing the process $X^\sim_t$ and of course $\nu^\sim(D) < +\infty$ for every open set D. Thus the above results are again applicable. We summarize in

Theorem 8.5. *If $(\underline{\underline{F}}, E)$ is irreducible and recurrent then the time changed Dirichlet space $(\underline{\underline{H}}, Q)$ is also irreducible and recurrent. Moreover $\underline{\underline{H}}_{(e)} = \gamma\underline{\underline{F}}_{(e)}$ and (8.4) is valid for φ in $\gamma\underline{\underline{F}}_{(e)}$.* ///

In the remainder of this section we use Theorems 8.2 and 8.5 to refine results from earlier sections.

To begin with, the second remark following Theorem 1.9 can now be extended to

Theorem 8.6. *If $(\underline{\underline{F}}, E)$ is recurrent, then 1 is in $\underline{\underline{F}}_{(e)}$ and $E(1,1) = 0$.* ///

8.2. Definition. Borel $f \geq 0$ is excessive if for quasi-every x the numbers $P_t f(x)$ decrease with t and $\lim_{t \downarrow 0} P_t f(x) = f(x)$. ///

Theorem 8.7. *Assume that $(\underline{\underline{F}}, E)$ is transient. If f is excessive and if there exists a potential g in $\underline{\underline{F}}_{(e)}$ such that $0 \leq f \leq g$ quasi-everywhere, then also f is a potential in $\underline{\underline{F}}_{(e)}$.* ///

8.11

Proof. Choose ν in $\mathfrak{M}$ equivalent to dx and such that g is in $L^2(\nu)$. Then g belongs to the time changed Dirichlet space $(\underline{F}^{\sim}, E)$. Since f is dominated by g it follows from the third remark following Theorem 1.9 that $P_t f \to 0$ quasi-everywhere as $t \uparrow \infty$ and therefore

$$G(1/s)\,(1-P_s)f$$

$$= \operatorname{Lim}_{T\uparrow\infty} \{(1/s)\int_0^s dt\, P_t f - (1/s)\int_T^{T+s} dt\, P_t f\}$$

$$= (1/s)\int_0^s dt\, P_t f$$

increases to f quasi-everywhere as $s \downarrow 0$. But for fixed $t > 0$ and for quasi-every x

$$P_t^{\sim} f(x) = \mathscr{E}_x f(X_{b(\nu;t)})$$

$$= \operatorname{Lim}_{s\downarrow 0} \mathscr{E}_x\, G(1/s)\,(1-P_s)\, f(X_{b(\nu;t)})$$

$$= \operatorname{Lim}_{s\downarrow 0} \mathscr{E}_x \int_{b(\nu;t)}^{\zeta} du (1/s)(1-P_s) f(X_u)$$

$$\leq \operatorname{Lim}_{s\downarrow 0} G(1/s)\,(1-P_s)\, f(x)$$

$$= f(x)$$

and it follows from Lemma 3.3 that possibly modulo quasi-continuous refinements f is a potential in $\underline{F}^{\sim}$ and therefore in $\underline{F}_{(e)}$. But since $P_t f \to f$ quasi-everywhere it follows from Lemma 4.6 that actually f is in $\underline{F}_{(e)}$. ///

The following is a prerequisite for showing in the recurrent case that functions in $\underline{F}_{(e)}$ have quasi-continuous refinements.

Lemma 8.8. Let ν in $\mathcal{M}_1$ be quivalent to dx and let $(\underline{\underline{F}}^{\sim}, E)$ be the corresponding time changed Dirichlet space. Then a set A is polar for $(\underline{\underline{F}}^{\sim}, E)$ if and only if it is polar for $(\underline{\underline{F}}, E)$ and a function f is quasi-continuous for $(\underline{\underline{F}}^{\sim}, E)$ if and only if it is quasi-continuous for $(\underline{\underline{F}}, E)$. ///

Proof. The transient case follows directly from Propositions 3.20 and 3.22. The recurrent case is easily handled by adapting the proof of those propositions with the help of Proposition 3.21. ///

Theorem 8.9. Assume that $(\underline{\underline{F}}, E)$ is recurrent.

(i) Every f in $\underline{\underline{F}}_{(e)}$ has a quasi-continuous refinement which is unique up to quasi-equivalence.

(ii) Let $f_n, n \geq 1$ and f be in $\underline{\underline{F}}_{(e)}$. Assume that $f_n \to f$ almost everywhere and that $E(f-f_n, f-f_n) \to 0$. Then there exists a subsequence such that for the quasi-continuous refinements, $f_n \to f$ quasi-everywhere. ///

Proof. For (i) it suffices to choose ν in $\mathcal{M}_1$ equivalent to dx such that f belongs to $L^2(\nu)$ and to apply Lemma 8.7 for existence and Lemma 3.16 for uniqueness. For (ii) it is easy to reduce to the special case when f_n and f are uniformly bounded by considering truncations and this can be handled by looking at the time changed Dirichlet space which corresponds to bounded ν in $\mathcal{M}_1$ equivalent to dx. ///

8.3. Convention. Unless otherwise specified every f in $\underline{\underline{F}}_{(e)}$ is represented by its quasi-continuous version. ///

This convention has already been established in Section 3 for the transient case.

8.13

Theorem 8.10. Let $\{f_n\}_{n=1}^{\infty}$ be a sequence in $F_{(e)}$ which is bounded in E norm.

(i) If (F, E) is transient then there exists f in $F_{(e)}$ such that the Cesàro sums of a subequence converge quasi-everywhere to f.

(ii) If (F, E) is recurrent and if the f_n are uniformly bounded, then there exists f in $F_{(e)}$ such that the Cesàro sums of a subsequence converge quasi-everywhere to f. ///

Proof. The transient case follows by the paragraph following 1.6.1' since $F_{(e)}$ is a Hilbert space. The recurrent case with dx bounded follows in the same way since the f_n are bounded also in E_1 norm and the general recurrent case then follows with the help of Theorem 8.5. ///

Chapter II. Decomposition of the Dirichlet Form

To our knowledge the main results appear here for the first time.

After some preliminaries in Section 9 the killing measure $\kappa(dx)$ and Lévy kernel $J(dx,dw)$ are defined in Section 10. These measure respectively the intensity for jumping to the dead point and the intensity for jumping within the state space. In Section 12 we give a direct characterization of κ and J. This can be used for example to obtain the main results of [45] without the side conditions imposed there. κ and J together determine a part of the Dirichlet form. The remainder we call the "Diffusion form" and in Section 11 we show that it is contractive and also local in the sense of [25].

9. Potentials in the Wide Sense.

In this section $(\underline{F}, E)$ is transient.

9.1. Definition. $f \geq 0$ is a potential in the wide sense if

9.1.1. f is defined and finite quasi-everywhere.

9.1.2. There exists a Radon measure μ charging no polar set such that $f = N\mu$. ///

In paragraph 3.11 we discussed the extension of the potential operator N to Radon measure μ charging no polar set. We will denote by $\mathcal{M}_{(e)}$ the set of such μ such that $N\mu$ is a potential in the wide sense.

We begin with the elementary

Lemma 9.1. (i) If μ is a Radon measure charging no polar set then either μ is in $\mathcal{M}_{(e)}$ or $N\mu = +\infty$ quasi-everywhere.

(ii) $\mathcal{M}_{(e)}$ contains every bounded Radon measure which charges no polar set. ///

Proof. Let $A = \{x : N\mu(x) = +\infty\}$. To prove (i) note that the process $\{N\mu(X_t), t \geq 0\}$ is an increasing limit of right continuous supermartingales and therefore

$$\mathcal{P}_x[\sigma(A) = +\infty] = 1 \tag{9.1}$$

for quasi-every x in $\underline{X}-A$. By irreducibility either A or $\underline{X}-A$ is dx null. If A is null then it is polar by Proposition 4.15. If $\underline{X}-A$ is null then Corollary 4.7-(i) together with (9.1) imply that $\underline{X}-A$ is polar. For (ii) it suffices to observe that if A were nonpolar then by Corollary 3.12 and the maximum principle Corollary 2.15 there exists ν in $\mathcal{M}$ with $\nu(A) > 0$ and $N\nu$ bounded. But then

$$\int \nu(dx)\, N\mu(x) = \int \mu(dx)\, N\nu(x) \leq \| N\nu \|_\infty \, \mu(\underline{X})$$

which is impossible. ///

Lemma 9.1 -(ii) is a natural refinement of the very definition of transience in Section 1. Lemma 9.1-(i) is valid with $N\mu$ replaced by general excessive f since any such f can be approximated from below by potentials in $\underline{\underline{F}}_{(e)}$.

Theorem 9.2. If f is a potential in the wide sense, then the measure μ of 9.2.2. is unique. ///

Proof. We can apply Theorem 8.2 and restrict attention to the special case where μ is dominated by dx. For $\varphi \geq 0$ such that both φ and φf are in $L^1(dx)$ we can then apply Lemma 1.3 to get

$$\begin{aligned}\int \mu(dx)\, \varphi(x) &= \operatorname{Lim}_{u \uparrow \infty} u \int \mu(dx)\, G_u\varphi(x) \\ &= \operatorname{Lim}_{u \uparrow \infty} u \int \mu(dx) \{ G\varphi(x) - GG_u\varphi(x) \} \\ &= \operatorname{Lim}_{u \uparrow \infty} u \int dx\, f(x) \{ \varphi(x) - uG_u\varphi(x) \}\end{aligned}$$

and the theorem follows since every finite nonnegative function can be approximated from below by such φ. ///

Finally we establish a useful sufficient condition for a given function to be a potential in the wide sense. Our proof is a simple variant of the now standard argument as given for example in [2, p. 268].

Theorem 9.3. *Let $h \geq 0$ be finite quasi-everywhere. Then the following conditions are sufficient for h to be a potential in the wide sense.*

9.2.1. *h is excessive.*

9.2.2. *$H^{\sim n}h \to 0$ almost everywhere as $n \uparrow \infty$.*

9.2.3. *For each n there exists a potential g_n such that $h \leq g_n$ quasi-everywhere on D_n.* ///

See Convention 5.1 for the meaning of D_n and $H^{\sim n}$.

Remark. The potential kernel for Brownian motion in $R^d, d \geq 3$ satisfies 9.2.1. and 9.2.2 but is not a potential in the wide sense since singleton sets are polar. Thus these two conditions by themselves are not sufficient in general. ///

For the proof note that by Theorem 8.7 $\min(h, g_n)$ has a quasi-continuous refinement h_n which is a potential. Put $h_n = N\mu_n$ and denote by $\mu_n^{(m)}$ the restriction of μ_n to D_m. The estimate

$$\begin{aligned} E(\mu_n^{(m)}) &\leq \int_{D_m} \mu_n(dx)\, h_n(x) \\ &\leq \int \mu_n(dx)\, H^m h_m(x) \\ &= E(h_n, H^m h_m) \\ &= \int \mu_m(dx)\, H^m h_n(x) \\ &\leq E(h_m, h_m) \end{aligned}$$

show that for each m the energy $E(\mu_n^{(m)})$ is bounded independent of n. After applying Lemma 3.4 and selecting a subsequence we can assume that $\mu_n \to \mu$ vaguely and that for each m there exists a measure $\mu^{(m)}$ in $\mathcal{M}$ such that $\mu^{(m)}$ dominates the restriction of μ to D_m and is dominated by the restriction of μ to $c\ell(D_m)$ and such that $N\mu_n^{(m)} \to N\mu^{(m)}$ weakly in $\underline{\underline{F}}_{(e)}$ as $n \uparrow \infty$. For $\varphi \geq 0$ in $\mathcal{M}^o$ having compact support and for each m

$$\int dx\, \varphi(x) h(x)$$

$$= \operatorname{Lim}_{n\uparrow\infty} \int dx\, \varphi(x)\, N\mu_n(x)$$

$$= \operatorname{Lim}_{n\uparrow\infty} \left\{ \int \mu_n^{(m)}(dx)\, G\varphi(x) + \int_{M_m} \mu_n(dx)\, G\varphi(x) \right\}$$

$$= \int \mu^{(m)}(dx)\, G\varphi(x) + \operatorname{Lim}_{n\uparrow\infty} \int_{M_m} \mu_n(dx)\, G\varphi(x).$$

The first term increases as $m \uparrow \infty$ to

$$\int \mu(dx)\, G\varphi(x) = \int dx\, \varphi(x)\, N\mu(x).$$

The second term is dominated by

$$\operatorname{Lim\,sup}_{n\uparrow\infty} \int \Pi^{\sim m}\mu_n\,(dx)\, G\varphi(x)$$

$$= \operatorname{Lim\,sup}_{n\uparrow\infty} \int dx\, \varphi(x)\, N\Pi^{\sim m}\mu_n(x)$$

$$= \operatorname{Lim\,sup}_{n\uparrow\infty} \int dx\, \varphi(x)\, H^{\sim m}\, N\mu_n(x)$$

$$\leq \int dx\, \varphi(x)\, H^{\sim m}\, h(x)$$

which decreases to 0 as $m \uparrow \infty$ by 9.2.2. Thus $h = N\mu$ almost everywhere and from 9.2.1 it follows that actually this equality holds quasi-everywhere.///

<u>Remark</u>. Actually $h_n = \min(h, g_n)$ quasi-everywhere. This follows from a result of Doob and Hunt [2, p. 285] concerning continuity of excessive functions along sample trajectories.

10. The Lévy Kernel

The functions

$$p(x) = \mathcal{P}_x[\zeta = +\infty]$$

$$r(x) = \mathcal{P}_x[\zeta < +\infty\ ;\ X_{\zeta - 0} \neq \partial]$$

$$s(x) = \mathcal{P}_x[\zeta < +\infty;\ X_{\zeta - 0} = \partial]$$

are defined quasi-everywhere. In the recurrent case $p=1$ and $r=s=0$. In the transient case

(10.1) $$1 = p + r + s.$$

For the moment we restrict attention to the transient case. It follows immediately from Theorem 9.3 that r is a potential in the wide sense and so

(10.2) $$r = N\kappa$$

with κ in $\mathcal{M}_{(e)}$. We will refer to κ as killing measure. Its significance is established in

Theorem 10.1. *Assume* $(\underline{F}, E)$ *is transient. Then for* $\varphi \geq 0$ *on* $\underline{X}$.

(10.3) $$\mathcal{E}_x[\zeta < +\infty;\ X_{\zeta-0} \neq \partial\ ;\ \varphi(X_{\zeta-0})] = N(\varphi\cdot\kappa)(x).$$

Also for $u > 0$

(10.3') $$\mathcal{E}_x[X_{\zeta-0} \neq \partial\ ;\ e^{-u\zeta}\,\varphi(X_{\zeta-0})] = N_u(\varphi\cdot\kappa)(x).\ ///$$

Proof. It suffices to prove (10.3') since (10.3) then follows after passage to the limit $u \downarrow 0$. For this purpose let R_u be the usual terminal time which is independent of the trajectory variables and which is exponentially distributed with density $ue^{-u\ell}$. Let $\zeta_u = \min(R_u, \zeta)$. Clearly

$$\mathcal{E}_x[R_u < \zeta ; \varphi(X_{R_u -0})] = u\, G_u \varphi$$

and so it suffices to establish

$$(10.4) \qquad \mathcal{E}_x[X_{\zeta_u -0} \neq \partial ; \varphi(X_{\zeta_u -0})] = N_u(\varphi \cdot \varkappa_u)(x)$$

where $\varkappa_u = \varkappa + u\,dx$. Of course it suffices to consider φ in $C_{com}(\underline{X})$ and since both sides of (10.4) are excessive it suffices to show that

$$(10.5) \qquad \int dx\, g(x)\, \mathcal{E}_x[X_{\zeta_u -0} \neq \partial ; \varphi(X_{\zeta_u -0})]$$

$$= \int \varkappa_u(dy)\, \varphi(y)\, G_u g(y)$$

for $g \geq 0$ in $C_{com}(\underline{X})$. Let

$$r_u(x) = \mathcal{P}_x[X_{\zeta_u -0} \neq \partial]$$
$$s_u(x) = \mathcal{P}_x[X_{\zeta_u -0} = \partial] = \mathcal{E}_x[X_{\zeta -0} = \partial ; e^{-u\zeta}]$$

and note that

$$1 = r_u + s_u$$

$$s = s_u + u\, G_u\, s$$

and therefore

$$(10.6) \qquad r_u = 1 - s_u = 1 - s + u\, G_u\, s = r + p + u\, G_u s$$

$$= N\varkappa + u\, G_u(p+s) = N\varkappa + u\, G_u\, 1 - uG_u r$$

$$= N\varkappa - u\, G_u\, N\varkappa + u\, G_u 1 = N_u \varkappa + u\, G_u 1$$

$$= N_u \varkappa_u .$$

Thus the left side of (10. 5)

$$= \operatorname{Lim}_{n\uparrow\infty} \Sigma_{k=0}^{\infty} \int dx\, g(x)\, \mathcal{E}_x[X_{\zeta_u - 0} \neq \partial\, ;\, \varphi\,(X_{k/2^n})\, ;\, k/2^n < \zeta_u \leq (k+1)/2^n\,]$$

$$= \operatorname{Lim}_{n\uparrow\infty} \Sigma_{k=0}^{\infty} \int dx\, g(x)\, e^{-uk/2^n}\, P_{k/2^n}\, \varphi\,\{\, r_u - e^{-u/2^n} P_{1/2^n} r_u \}\,(x)$$

$$= \operatorname{Lim}_{n\uparrow\infty} \Sigma_{k=0}^{\infty} \int \kappa_u(dy)\, G_u(1-e^{-u/2^n} P_{1/2^n})\,\varphi\, e^{-uk/2^n}\, P_{k/2^n}\, g(y)$$

$$= \operatorname{Lim}_{n\uparrow\infty} \Sigma_{k=0}^{\infty} \int \kappa_u(dy) \int_0^{1/2^n} dt\, e^{-ut}\, P_t \varphi\, e^{-uk/2^n}\, P_{k/2^n}\, g(y).$$

The theorem follows since

$$P_t\, \varphi\, P_{h-t}\, g(y) = \mathcal{E}_y\, \varphi(X_t) g(X_h)$$

converges to $\varphi(y)\, P_h g(y)$ as $t \downarrow 0$ for all $h > 0$ and for quasi-every y. ///

In the remainder of the section we allow $(\underline{F}, E)$ to be either transient or recurrent.

Lemma 10.2. *Let D be an open subset of $\underline{X}$ such that $M = \underline{X} - D$ is nonpolar.*

(i) *The absorbed Dirichlet space $(\underline{F}^D, E)$ is regular and transient.*

(ii) *A subset A of D is polar for $(\underline{F}^D, E)$ if and only if it is polar for $(\underline{F}, E)$.* ///

Proof. Transience is obvious when $(\underline{F}, E)$ is transient and was noted in Lemma 8.3 when $(\underline{F}, E)$ is recurrent. To show that $\underline{F}^D \cap C_{com}(D)$ is uniformly dense in $C_{com}(D)$ fix $\varphi \geq 0$ in $C_{com}(D)$ and $\epsilon > 0$ and choose $f \geq 0$ in $\underline{F} \cap C_{com}(X)$ such that $\|\varphi - f\|_\infty < \epsilon$. Then $g = f - \min(f, \epsilon)$ belongs to $\underline{F}^D \cap C_{com}(D)$ and $\|\varphi - g\|_\infty < 2\epsilon$. In showing that $\underline{F}^D \cap C_{com}(D)$

is dense in $\underline{\underline{F}}^D$ it suffices to approximate bounded nonegative f in $\underline{\underline{F}}^{D'}$ where D' is open and has compact closure contained in D. By the above result there exists $\varphi \geq 0$ in $\underline{\underline{F}}^D \cap C_{com}(D)$ such that $\varphi \geq f$ quasi-everywhere. Choose nonnegative g_n in $\underline{\underline{F}} \cap C_{com}(X)$ such that $g_n \to f$ quasi-everywhere and and $E_1(g_n, g_n)$ is bounded. Then the functions $\min(\varphi, g_n)$ belong to $\underline{\underline{F}}^D \cap C_{com}(D)$, converge to f quasi-everywhere, and are bounded in E_1 norm. Therefore the Cesàro sums of a subsequence converge to f in $\underline{\underline{F}}^D$ and regularity of $(\underline{\underline{F}}^D, E)$ is established. If a subset A of D is polar for $(\underline{\underline{F}}, E)$ then because of Proposition 4.15 it is probabilistically obvious that A is polar for $(\underline{\underline{F}}^D, E)$. The converse follows since the absorbed capacity Cap_1^D dominates Cap_1 and the lemma is completely proved.///

Fix D and M as in Lemma 10.2 and for x in D define

$$(10.7) \qquad p^D(x) = \mathcal{P}_x[\zeta = +\infty\ ; \sigma(M) = +\infty]$$

$$r^D(x) = \mathcal{P}_x[\zeta < +\infty\ ; \sigma(M) = +\infty; X_{\zeta-0} \neq \partial]$$

$$s^D(x) = \mathcal{P}_x[\zeta < +\infty; \sigma(M) = +\infty; X_{\zeta-0} = \partial]$$

$$h_d^D(x) = \mathcal{P}_x[\sigma(M) < +\infty; X_{\sigma(M)-0} \in M]$$

$$h_j^D(x) = \mathcal{P}_x[\sigma(M) < +\infty; X_{\sigma(M)-0} \in D].$$

It follows from Proposition 4.16 that if (F, E) is recurrent and if M is nonpolar then $p^D = r^D = s^D = 0$. If D has compact closure, then **also** $p^D = s^D = 0$ in the transient case.

<u>Theorem 10.3</u>. <u>Let</u> D <u>be open and assume that</u> $M = \underline{\underline{X}} - D$ <u>is nonpolar</u>.

(i) <u>In the transient case</u> $r^D = N^D\kappa$. <u>Also for</u> $\varphi \geq 0$ <u>on</u> D.

$$(10.8) \qquad \mathcal{E}_x[\zeta < \sigma(M); X_{\zeta-0} \neq \partial; \varphi(X_{\zeta-0})] = N^D(\varphi \cdot \kappa)(x)$$

<u>and for</u> $u > 0$

(10.8') $\mathcal{E}_x[e^{-u\zeta}; \zeta < \sigma(M); X_{\zeta-0} \neq \partial; \varphi(X_{\zeta-0})] = N_u^D(\varphi\cdot\kappa)(x).$

(ii) <u>In the transient case</u>

$$p^D = uG_u^D p^D$$
$$r^D = uG_u^D r^D + N_u^D \kappa$$
$$s^D = uG_u^D s^D + \mathcal{E}_x[\sigma(M) = +\infty; X_{\zeta-0} = \partial; e^{-u\zeta}]$$

(iii) <u>In either the transient or recurrent case there exists a unique</u> Radon measure κ^D <u>on</u> D <u>charging no polar set such that</u>

$$h_j^D = H^D \kappa^D . ///$$

<u>Proof.</u> The first sentence in (i) follows from (10.2) and (7.3) since clearly $r^D = r - H^M r$. The remainder of (i) follows by the proof of Theorem 10.1 Conclusion (ii) follows by direct computation and (iii) follows upon applying Theorem 9.3 to h_j^D with $(\underline{E}^D, E)$ playing the role of $(\underline{E}, E)$.///

We continue to work with D and M as in Theorem 10.3. It is easy to see that for each nonnegative φ in $C_{com}(\underline{X})$ there exists a function $J^D(\cdot, \varphi)$ specified up to κ^D equivalence such that

(10.9) $\mathcal{E}_x[\sigma(M) < \zeta; X_{\sigma(M)-0} \in D; \varphi(X_{\sigma(M)})]$

$$= N^D \kappa^D J^D(\cdot, \varphi)(x)$$

for quasi-every x in D. Letting φ vary in the usual manner, we establish the existence of a kernel $J^D(y, dz)$ such that

(10.10) $\mathcal{E}_x[\sigma(M) < \zeta; X_{\sigma(M)-0} \in D; \varphi(X_{\sigma(M)})]$

$$= N^D\{\kappa^D \cdot \int J^D(\cdot, dz)\varphi(z)\}(x)$$

for $\varphi \geq 0$ on M. (The precise technical conditions on J^D are that $J^D(y, \cdot)$ is a Borel probability [a. e. $\kappa(dy)$] and that for every Borel set Γ the function $J^D(\cdot, \Gamma)$ is Borel measurable.) The proof of Theorem 10.1 is easily adapted to refine (10.10) to

(10.11) $$\mathcal{E}_x[\sigma(M) < +\infty;\ X_{\sigma(M)-0} \in D;\ \psi(X_{\sigma(M)-0})\,\varphi(X_{\sigma(M)})]$$

$$= N^D\{\varkappa^D \cdot \psi \int J^D(\cdot\,,\ dz)\,\varphi(z)\}(x)$$

and at the same time establish the $u > 0$ version

(10.11') $$\mathcal{E}_x[X_{\sigma(M)-0} \in D;\ e^{-u\sigma(M)}\,\psi(X_{\sigma(M)-0})\,\varphi(X_{\sigma(M)})]$$

$$= N_u^D\{\varkappa^D \cdot \psi \int J^D(\cdot\,,\ dz)\,\varphi(z)\}(x)$$

for $\varphi \geq 0$ on M and $\psi \geq 0$ on D.

Lemma 10.4. (i) If D, D' are open sets with $D' \subset D$ then

(10.12) $$\int \varkappa^{D'}(dy)\ \psi(y) \int J^{D'}(y, dz)\,\varphi(z)$$

$$= \int_{D'} \varkappa^D(dy)\ \psi(y) \int J^D(y, dz)\,\varphi(z)$$

for $\psi \geq 0$ on D' and $\varphi \geq 0$ on $M = \underline{\underline{X}} - D$.

(ii) For $u > 0$ and for quasi-every x

(10.13) $$\mathcal{E}_x \sum_t e^{-ut} I(X_{t-0} \in D;\ X_t \in M)\,\psi(X_{t-0})\,\varphi(X_t)$$

$$= N_u\{\varkappa^D \cdot \psi \int J^D(\cdot\,,\ dz)\,\varphi(z)\}(x)$$

for $\psi \geq 0$ on D and $\varphi \geq 0$ on $M = \underline{\underline{X}} - D$.

(iii) For $u > 0$ let $\mathcal{E}_{(u)}$ be the approximate Markov process of Section 5 which corresponds to $(\underline{F}, E_u)$. Then

$$(10.14) \quad \int \varkappa^D(dy)\, \psi(y) \int J^D(y, dz)\, \varphi(z)$$

$$= \mathcal{E}_{(u)} \Sigma_t I(X_{t-0} \in D\,;\, X_t \in M)\, \psi(X_{t-0})\, \varphi(X_t)$$

for $\psi \geq 0$ on D and $\varphi \geq 0$ on M.

(iv) If D, D' are disjoint open sets then

$$(10.15) \quad \int \varkappa^D(dy)\, \psi(y) \int J^D(y, dz)\, \varphi(z)$$

$$= \int \varkappa^{D'}(dy)\, \varphi(y) \int J^{D'}(y, dz)\, \psi(z)$$

for $\psi \geq 0$ on D and $\varphi \geq 0$ on D'. ///

Note. Of course (10.13) and (10.14) are valid also for $u = 0$ when $(\underline{F}, E)$ is transient.

Proof. To prove (10.12) define g on D and g' on D' by

$$g(x) = \mathcal{E}_x [\sigma(M) < \zeta\,;\, X_{\sigma(M)-0} \in D'\,;\, \psi(X_{\sigma(M)-0})\, \varphi(X_{\sigma(M)})]$$

$$g'(x) = \mathcal{E}_x [\sigma(M) < \zeta\,;\, X_{\sigma(M')-0} \in D'\,;\, X_{\sigma(M')} \in M;\, \psi(X_{\sigma(M')-0})\varphi(X_{\sigma(M')})].$$

By (10.11)

$$g = N^D \{\varkappa^D \cdot \psi\, 1_{D'} \int J^D(\cdot, dz)\, \varphi(z)\}$$

$$g' = N^{D'} \{\varkappa^{D'} \cdot \psi \int J^{D'}(\cdot, dz)\, 1_M(z)\, \varphi(z)\}.$$

From the definition of g and g' clearly

$$g = g' + H^{M'}(1_D g).$$

On the other hand by an appropriate relative version of (7.3)

$$g = N^{D'}\{\varkappa^D \cdot \psi 1_{D'} \int J^D(\cdot, dz)\varphi(z)\} + H^{M'} 1_D g$$

and (10.12) follows from the uniqueness result Theorem 9.2 applied to $(\underline{F}^{D'}, E)$. In (10.13) it suffices to consider the special case when φ, ψ are bounded with disjoint compact supports. Fix $u > 0$ and let f, g be respectively the left and right sides of (10.13). To identify f and g let D' be open with compact support contained in D and such that D' contains the support of φ. Let

$$\sigma^{\sim}(M) = \inf\{t > \sigma(D') : X_t \in M\}$$

and let $\nu = \varkappa^D \cdot \psi \int J^D(\cdot, dz)\varphi(z)$. For x in $\underline{X}$ define

$$(10.16) \qquad f_0(x) = \mathcal{E}_x[X_{\sigma^{\sim}(M)-0} \in D; e^{-u\sigma^{\sim}(M)}\psi(X_{\sigma^{\sim}(M)-0})\varphi(X_{\sigma^{\sim}(M)})]$$

$$g_0(x) = \mathcal{E}_x \int_0^{\sigma^{\sim}(M)} a(\nu; dt)\, e^{-ut}.$$

It is easy to check that f, g are the minimal nonnegative solutions of

$$(10.17) \qquad f(x) = f_0(x) + \mathcal{E}_x e^{-u\sigma^{\sim}(M)} f(X_{\sigma^{\sim}(M)})$$

$$g(x) = g_0(x) + \mathcal{E}_x e^{-u\sigma^{\sim}(M)} g(X_{\sigma^{\sim}(M)})$$

and so (10.13) will be established if we show that $f_0 = g_0$. (This reduction would be invalid if we replaced $\sigma^{\sim}(M)$ by $\sigma(M)$ in (10.16 and (10.17).)
But clearly

$$f_0(x) = H_u^{D'}\{\mathcal{E}_\cdot\,[X_{\sigma(M)-0}\in D\,;\, e^{-u\sigma(M)}\psi(X_{\sigma(M)-0})\varphi(X_{\sigma(M)})]\}(x)$$

$$g_0(x) = H_u^{D'}\{\mathcal{E}_\cdot \int_0^{\sigma(M)} a(\nu;\, dt)\, e^{-ut}\}(x)$$

$$= H_u^{D'} N_u^D \nu(x)$$

and the desired equality follows from (10.11'). To prove (10.14) let $L_{(u)k}$ be the unique measure in $\mathcal{M}_u$ such that

$$H_u^k 1 = N_u L_{(u)k} .$$

Then the right side of (10.14)

$$= \mathrm{Lim}_{k\uparrow\infty} \int L_{(u)k}(dx)\, \mathcal{E}_x \Sigma_t e^{-ut} I(X_{t-0}\in D\,;\, X_t\in M)\psi(X_{t-0})\varphi(X_t)$$

$$= \mathrm{Lim}_{k\uparrow\infty} \int L_{(u)k}(dx)\, N_u\{\varkappa^D\cdot \psi\int J^D(\cdot, dz)\varphi(z)\}(x)$$

$$= \mathrm{Lim}_{k\uparrow\infty} \int \varkappa^D(dy)\psi(y)\, H_u^k 1(y) \int J^D(y, dz)\varphi(z)$$

$$= \int \varkappa^D(dy)\, \psi(y) \int J^D(y, dz)\, \varphi(z)$$

and (10.14) is proved. Finally (10.15) follows from (10.14) and invariance under time reversal, Theorem 5.3. ///

From (10.12) and (10.15) it follows that there exists a unique symmetric measure $J(dy, dz)$ on $\underline{\underline{X}} \times \underline{\underline{X}}$ such that

$$(10.18) \qquad \varkappa^D(dy)\, J^D(y, dz) = 1_D(y) J(dy, dz)\, 1_M(z)$$

for any choice of D as above. We will refer to $J(dy, dz)$ as the <u>Lévy kernel</u> for the Dirichlet space. A routine passage to the limit in (10.14) yields

Theorem 10.5. For $u > 0$ and for $\varphi, \psi \geq 0$ on $\underline{X}$

$$\mathcal{E}_{(u)} \Sigma_t I(X_t \neq X_{t-0};\ X_t, X_{t-0} \in \underline{X}) \psi(X_{t-0}) \varphi(X_t) \tag{10.19}$$

$$= \iint J(dy, dz)\ \psi(y)\ \varphi(z).$$

where $\mathcal{E}_{(u)}$ is the approximate Markov process of Section 5 which corresponds to $(\underline{F}, E_u)$. This is also true for $u = 0$ when $(\underline{F}, E)$ is transient. ///

Clearly $J(dy, dz)$ is the same as the measure $\sigma(dx, dy)$ of [1, p. 212]. Also it corresponds to the Lévy measures of S. Watanabe in [50].

11. The Diffusion Form

To simplify the presentation, we assume throughout this section that $(\underline{F}, E)$ is transient. However theorems which are stated without the qualification that $(\underline{F}, E)$ be transient are true also in the recurrent case, although the proof will then be given only for the transient case, with the recurrent case being taken for granted or mentioned in a remark.

Lemma 11.1. *Assume that $(\underline{F}, E)$ is transient. Then*

(11.1) $$\mathrm{Lim}_{t\downarrow 0}(1/t)\int dx\,\{1-P_t 1(x)\}\,\varphi(x) = \int \varkappa(dx)\,\varphi(x)$$

(11.1') $$\mathrm{Lim}_{u\downarrow\infty}\, u\int dx\,\{1-u\,G_u 1(x)\}\,\varphi(x) = \int \varkappa(dx)\,\varphi(x)$$

if either $\varphi \in C_{com}(\underline{X})$ or if $\varphi \in \underline{F}$ and also φ is bounded with compact support. In particular $(1/t)\{1-P_t 1(x)\}dx$ as $t\uparrow\infty$ and $u\{1-uG_u 1(x)\}dx$ as $u\uparrow\infty$ converge vaguely to $\varkappa(dx)$ ///

Proof. It suffices to prove (11.1) since

$$u\,\{1-u\,G_u 1(x)\}$$

$$= \int_0^\infty dt\; u^2 te^{-ut}\,(1/t)\,\{1-P_t 1(x)\}.$$

But for φ as specified in the lemma

(11.2) $$\mathrm{Lim}_{t\downarrow 0}\; I(\zeta<+\infty)\,(1/t)\int_{\zeta-t}^{\zeta} dx\,\varphi(X_s)$$

$$= I(\zeta<+\infty;\; X_{\zeta-0}\neq\partial)\,\varphi(X_{\zeta-0})$$

[a.e. θ] and also the left side is uniformly bounded and supported by a set of finite θ measure Therefore (11.1) follows from (10.3) and the calculation

$$(11.3) \qquad \int dx\, (1/t)\{1-P_t 1(x)\}\ \varphi(x)$$

$$= (1/t)\mathcal{E}\int_{\zeta^*}^{\zeta} ds\ \varphi(X_s)\{1-P_t 1(X_t)\}$$

$$= (1/t)\ \mathcal{E}\ \int_{\zeta^*}^{\zeta} ds\ \varphi(X_s)\ I(s < \zeta \leq s+t)$$

$$= \mathcal{E}\, I(\zeta < +\infty)\ (1/t)\int_{\zeta - t}^{\zeta} ds\ \varphi(X_s).\ ///$$

Corollary 11.2. Assume that (F, E) is transient.

(i) If f belongs to $F_{(e)}$ then

$$(11.4) \qquad \mathrm{Lim}_{t \downarrow 0}(1/t)\int dx\ \{1-P_t 1\ (x)\}\ f^2(x) = \int \varkappa(dx)\ f^2(x)$$

(ii) If f, g belong to $\underline{\underline{F}}_{(e)}$ and are bounded, then

$$(11.5) \qquad \mathrm{Lim}_{t \downarrow 0}\ (1/t)\ \int dx\ \{1-P_t 1(x)\}\ \ g(x)\ f^2(x) = \int \varkappa(dx)\ g(x) f^2(x).\ ///$$

Proof. (ii) follows from (i) since then fg belongs to $\underline{\underline{F}}_{(e)}$ and (i) follows from Lemma 11.1 and the usual "3-ε argument" since

$$(11.6) \qquad (1/t)\int dx\ \{1-P_t 1\}\ (x)\ f^2(x)$$

$$\leq (1/t)\int dx\ \{1-P_t 1\ (x)\}\ f^2(x)$$

$$+ (1/t)\int dx \int P_t(x, dy)\{f(x)-f(y)\}^2$$

$$\leq E(f, f)$$

by Lemma 1.7. ///

Remark. The "Laplace transform" versions of (11.4) and (11.5) are also valid. The same is true for other relations derived below. ///

For $t > 0$ define

(11.7) $$\langle A_t f\rangle(x) = (1/t)\int P_t(x,dy)\{f(x)-f(y)\}^2$$

when it converges. Of course (11.7) converges quasi-everywhere for f in $\underline{\underline{F}}_{(e)}$.

Theorem 11.3. *Let f in $\underline{\underline{F}}_{(e)}$ be bounded. Then there exists a unique Radon measure $\langle Af\rangle(dx)$ charging no polar set such that*

(11.8) $$\mathrm{Lim}_{t\downarrow 0}\int dx\,\langle A_t f\rangle(x)g(x) = \int\langle Af\rangle(dx)\,g(x)$$

for bounded g in $\underline{\underline{F}}_{(e)}$. In particular $\mathrm{Lim}_{t\downarrow 0}\langle A_t f(x)\rangle dx = \langle Af\rangle(dx)$ vaguely. Moreover for bounded g in $\underline{\underline{F}}_{(e)}$

(11.9) $$\frac{1}{2}\int\langle Af\rangle(dx)\,g(x) + \frac{1}{2}\int\kappa(dx)\,f^2(x)g(x)$$
$$= E(gf,f) - \frac{1}{2}E(g,f^2).\ ///$$

Proof. Consider first g in $\underline{\underline{F}}_{(e)} \cap C_{com}(\underline{X})$. Then by symmetry and since individual terms converge separately

(11.10) $$(1/t)\int dx\int P_t(x,dy)[\{g(x)f(x)-g(y)f(y)\}\{f(x)-f(y)\} - \frac{1}{2}\{g(x)-g(y)\}\{f^2(x)-f^2(y)\}]$$
$$= (1/t)\int dx\int P_t(x,dy)\,[2g(x)f(x)\{f(x)-f(y)\} - g(x)\{f^2(x)-f^2(y)\}]$$
$$= (1/t)\int dx\int P_t(x,dy)g(x)\{f^2(x)-2f(x)f(y)+f^2(y)\}$$
$$= \int dx\,g(x)\,\langle A_t f\rangle(x)$$

and it follows from Lemma 1.7 that

(11.11) $$\mathrm{Lim}_{t\downarrow 0}\left[\frac{1}{2}\int dx\,\langle A_t f\rangle(x)g(x) + \frac{1}{2}(1/t)\int dx\{1-P_t 1(x)\}\,g(x)f^2(x)\right]$$
$$= E(gf,f) - \frac{1}{2}E(g,f^2).$$

From (11.11) and (11.5) follows the existence of a Radon measure $\langle Af\rangle(dx)$ such that $\langle A_t f\rangle(x)\,dx \to \langle Af\rangle(dx)$ vaguely and then (11.8) and (11.9) are valid for g in $\underline{\underline{F}}_{(e)} \cap C_{com}(\underline{\underline{X}})$. It only remains to extend (11.8) and (11.9) to general bounded g in $\underline{\underline{F}}_{(e)}$. The estimate

(11.12) $$\sup_{t>0} \tfrac{1}{2} \int dx \,\langle A_t f\rangle(x)\, g(x) \le \{ E(gf, gf) E(f,f)\}^{\frac{1}{2}} + \tfrac{1}{2}\{ E(g,g)\, E(f^2, f^2)\}^{\frac{1}{2}}$$

follows for g in $\underline{\underline{F}}_{(e)} \cap C_{com}(\underline{\underline{X}})$ from (11.10), from Lemma 1.7 and from the Cauchy-Schwarz inequality and follows for general bounded g in $\underline{\underline{F}}_{(e)}$ with the help of Fatou's lemma. We first use (11.12) to show that $\langle Af\rangle$ charges no polar set. Fix a compact polar set K and choose g_n in $\underline{\underline{F}}_{(e)}$ such that $0 \le g_n \le 1$, such that for each n we have $g_n = 1$ almost everywhere on a neighorhood of K and such that $E(g_n, g_n) \to 0$. The products fg_n are bounded in E norm and a subsequence converges to 0 quasi-everywhere. Therefore after possibly replacing the g_n by Cesàro sums of a subsequence we can assume that also $E(fg_n, fg_n) \to 0$. It is still true that each $g_n = 1$ almost everywhere on a neighborhood of K and therefore

$$\langle Af\rangle(K) \le \inf_n \operatorname{Lim\,inf}_{t\downarrow 0} \int dx\,\langle A_t f\rangle(x)\, g_n(x) = 0.$$

Now fix bounded g in $\underline{\underline{F}}_{(e)}$ and choose uniformly bounded g_n in $\underline{\underline{F}}_{(e)} \cap C_{com}(\underline{\underline{X}})$ such that $g_n \to g$ and also $g_n f \to gf$ in $\underline{\underline{F}}_{(e)}$. Since $\langle Af\rangle$ charges no polar set we can assume also that $g_n \to g$ [a.e. $\langle Af\rangle$] and then it follows with the help of (11.12) that $\operatorname{Lim}_{n\uparrow\infty} \int \langle Af\rangle(dx)\, g_n(x) = \int \langle Af\rangle(dx) g(x)$. Finally another application of (11.12) shows that $\operatorname{Lim}_{t\downarrow 0} \int dx\,\langle A_t f\rangle(x)\, g(x)$ exists and is equal to $\text{Lim}_{n\uparrow\infty} \text{Lim}_{t\downarrow 0} \int dx \langle A_t f\rangle g_n(x)$ and the theorem follows. ///

Remark. The above proof requires only trivial modifications to be applicable to the recurrent case. ///

If f belongs to $\underline{\underline{F}}_{(e)}$ and is bounded and if f' is a normalized contraction of f then clearly

$$(11.13)\qquad \langle Af'\rangle(dx) \le \langle Af\rangle(dx).$$

For general f in $\underline{\underline{F}}_{(e)}$ let

$$f_n(x) = \begin{cases} n \text{ sgn } f(x) & \text{if } |f(x)| > n \\ f(x) & \text{if } |f(x)| \geq n. \end{cases}$$

and then define

$$(11.14)\qquad \langle Af\rangle(dx) = \text{Lim}_{n\uparrow\infty} \langle Af_n\rangle(dx).$$

This makes sense since by (11.13) the right side of (11.14) increases with n. Clearly the contractivity property (11.13) is valid for arbitrary f in $\underline{\underline{F}}_{(e)}$.

Theorem 11.4. Assume that $(\underline{\underline{F}}, E)$ is transient. Then for f in $\underline{\underline{F}}_{(e)}$ and for quasi-every x in X

$$(11.15)\qquad \mathcal{E}_x\{ Mf(\zeta) - Mf(0)\}^2 = N\langle Af\rangle(x) + N(f^2 . \kappa)(x)$$

Also if f is bounded and if ν is a bounded Radon measure charging no polar set such that $N\nu$ is bounded then

$$(11.16)\qquad \int \nu(dx)\, \mathcal{E}_x \{Mf(\zeta) - Mf(0)\}^2$$

$$= \text{Lim}_{t\downarrow 0} \int \nu(dx)\{G \langle A_t f\rangle (x) + G(1/t)\,(1-P_t 1) f^2(x)\} .$$

$$= \text{Lim}_{t\downarrow 0} \int \nu(dx)\, \mathcal{E}_x \int_0^\zeta ds\, (1/t)\{f(X_{s+t}) - f(X_s)\}^2 \quad ///$$

Proof. It suffices to establish (11.16) since then (11.15) will follow from (11.5) and (11.8) with the help of duality and a passage to the limit in f. As a preliminary for (11.16) we show that for the special case $f = G\varphi$ with φ bounded and for quasi-every x.

$$(11.17)\qquad \mathcal{E}_x\{Mf(\zeta) - Mf(0)\}^2 = \text{Lim}_{t\downarrow 0}\, (1/t)\, \mathcal{E}_x \int_0^\zeta ds\, \{f(X_{s+t}) - f(X_s)\}^2$$

From

$$(1/t)\, \mathcal{E}_x \int_0^\zeta ds\, \{Mf(s+t) - Mf(s)\}^2$$

$$= (1/t)\, \mathcal{E}_x \int_0^\zeta \langle Mf\rangle (dr) \int_{(r-t)^+}^{r} ds$$

it follows that

$$(11.18)\qquad \mathcal{E}_x \{Mf(\zeta) - Mf(0)\}^2 = \text{Lim}_{t\downarrow 0} (1/t)\, \mathcal{E}_x \int_0^\zeta ds\, \{Mf(s+t) - Mf(s)\}^2 .$$

and (11.17) will follow if we show that

$$(11.19)\qquad \text{Lim}_{t\downarrow 0}\, (1/t)\, \mathcal{E}_x \int_0^\zeta ds\, \{ \int_s^{s+t} dr\, \varphi(X_r)\}^2 = 0$$

$$(11.20)\qquad \text{Lim}_{t\downarrow 0}\, (1/t)\, \mathcal{E}_x \int_0^\zeta ds\, \{ f(X_{s+t}) - f(X_s)\} \int_s^{s+t} dr\, \varphi(X_r) = 0 .$$

But (11.19) follows from the estimate

$$(1/t)\mathcal{E}_x \int_0^\zeta ds\,\{\int_s^{s+t} dr\,\varphi(X_r)\}^2$$

$$\leq \|\varphi\|_\infty \mathcal{E}_x \int_0^\zeta ds \int_s^{s+t} dr\,\varphi(X_r)$$

$$\leq \|\varphi\|_\infty \mathcal{E}_x \int_0^\zeta dr\,\varphi(X_r)\,t$$

and then (11.20) follows from (11.19) and the Cauchy-Schwarz inequality since

$$(1/t)\,\mathcal{E}_x \int_0^\zeta ds\,\{f(X_{s+t})-f(X_s)\}^2$$

$$\leq 2\,(1/t)\,\mathcal{E}_x \int_0^\zeta ds\,\{Mf(s+t) - Mf(x)\}^2$$

$$+\,2(1/t)\,\mathcal{E}_x \int_0^\zeta ds\,\{\int_s^{s+t} dr\,\varphi(X_r)\}^2 .$$

With ν as specified

$$\int \nu(dx)\,\mathcal{E}_x \int_0^\zeta dr\,\varphi(X_r) \leq \nu(\underline{X})\,\|f\|_\infty$$

$$\int \nu(dx)\,(1/t)\,\mathcal{E}_x \int_0^\zeta ds\,\{f(X_{s+t}) - f(X_s)\}^2$$

$$\leq 2\{E(fN\nu, fN\nu)\,E(f,f)\}^{\frac{1}{2}} + \{E(N\nu, N\nu)\,E(f^2,f^2)\}^{\frac{1}{2}}$$

and so the above argument can be repeated to establish (11.16) for such f. Finally a passage to the limit in f is easily accomplished since for f, f' in $\underline{\underline{F}}_{(e)}$

$$|\int \nu(dx)\{G < A_t f'> (x) + G(1/t)(1-P_t 1)(f^2-f'^2)(x)\}|$$

$$\leq \{\int dx < A_t(f-f') > (x)\,N\nu(x) + \int dx(1/t)\{1-P_t 1(x)\}\,(f-f')^2(x)\,N\nu(x)\}^{\frac{1}{2}}$$

$$\times\ \{\int dx < A_t(f+f') > (x)\,N\nu(x) + \int dx(1/t)\{1-P_t 1(x)\}\,(f+f')^2(x)N\nu(x)\}^{\frac{1}{2}}$$

which is denominated by an expression which $\to 0$ as $(f-f')^2$, $(f-f')N\nu \to 0$ in $\underline{\underline{F}}_{(e)}$ with $(f+f')^2$, $(f+f')N\nu$ staying bounded in E norm and since a corresponding estimate holds for

$$\left| \int \nu(dx)\, \mathcal{E}_x \{Mf(\zeta) - Mf(0)\}^2 - \int \nu(dx)\, \mathcal{E}_x \{Mf'(\zeta) - Mf'(0)\}^2 \right|$$

whenever (11.15) is valid for f and f'. ///

Remark. For the recurrent case (11.15) and (11.16) should be replaced by

(11.15') $$\mathcal{E}_x \{Mf(R_u) - Mf(0)\}^2 = N_u \langle Af \rangle (x)$$

(11.16') $$\int \nu(dx)\, \mathcal{E}_x \{Mf(R_u) - Mf(0)\}^2$$

$$= \text{Lim}_{t \downarrow 0} \int \nu(dx)\, G_u \langle A_t f \rangle (x)$$

$$= \text{Lim}_{t \downarrow 0} \int \nu(dx)\, \mathcal{E}_x \int_0^{R_u} ds (1/t) \{f(X_{s+t}) - f(X_s)\}^2$$

where now $N_u \nu$ is bounded. ///

The identity (11.15) together with uniqueness in [35, VIII.3] and the simple Markov property leads immediately to

Theorem 11.5. For f in $\underline{\underline{F}}_{(e)}$

(11.21) $$\langle Mf \rangle (t) = a(\langle Af \rangle ; t) + a(f^2 \cdot \varkappa ; t) \quad ///$$

Next (11.21) and (11.13) give

Theorem 11.6. If f is in $\underline{\underline{F}}_{(e)}$ and if f' is a normalized contraction of f then

(11.22) $\langle Mf' \rangle (dt) \leq \langle Mf \rangle (dt)$. ///

11.1. Convention. Relations such as (11.21) and (11.22) are understood to be valid except for a set of sample paths in Ω_∞ which is null for $\mathcal{P}$ and a set in Ω which is null for $\mathcal{P}_x$ for quasi-every x. ///

Next we use (11.17) to establish the contractivity property (11.22) for the continuous part $\langle M_c f \rangle$. If ν satisfies the hypotheses of Theorem 11.4 then so does

$$\nu'(dy) = \int \nu(dx)\, \mathcal{E}_x [\tau < \zeta ;\ X_\tau \in dy].$$

for any stopping time τ and if follows that

$$\int \nu(dx)\, \mathcal{E}_x \{Mf(\zeta) - Mf(\tau)\}^2$$

$$= \mathrm{Lim}_{t \downarrow 0} \int \nu(dx)\, \mathcal{E}_x \int_\tau^\zeta ds\ (1/t)\ \{f(X_{s+t}) - f(X_s)\}^2$$

for bounded f in $\underline{\underline{F}}_{(e)}$ and therefore

$$\int \nu(dx)\, \mathcal{E}_x \{Mf(\tau) - Mf(0)\}^2$$

$$= \mathrm{Lim}_{t \downarrow 0} \int \nu(dx)\, \mathcal{E}_x \int_0^\tau ds\ (1/t)\ \{f(X_{s+t}) - f(X_s)\}^2 .$$

On the other hand

$$\mathrm{Lim}_{t \downarrow 0} \int \nu(dx)\, \mathcal{E}_x \int_0^\tau ds\ I(\tau \leq s + t)(1/t)\ \{f(X_{s+t}) - f(X_s)\}^2$$

$$= \mathrm{Lim}_{t \downarrow 0} \int \nu(dx)\, \mathcal{E}_x I(\tau < +\infty)\ (1/t) \int_{(\tau - t)^+}^\tau ds\ \{f(X_{s+t}) - f(X_s)\}^2$$

and therefore

(11.23) $$\int \nu(dx)\, \mathcal{E}_x (\{Mf(\tau) - Mf(0)\}^2 - I(\tau < +\infty)\ \{f(X_\tau) - f(x_{\tau - 0})\}^2)$$

$$= \mathrm{Lim}_{t \downarrow 0} \int \nu(dx)\, \mathcal{E}_x\ (1/t) \int_0^\tau ds\ I(s + t < \tau)\ \{f(X_{s+t}) - f(X_s)\}^2$$

In particular

(11.24) $$\mathscr{E}_x(\{Mf'(\tau) - Mf'(0)\}^2 - I(\tau < +\infty)\{f'(X_\tau) - f'(X_{\tau-0})\}^2)$$

$$\leq \mathscr{E}_x(\{Mf(\tau) - Mf(0)\}^2 - I(\tau < +\infty)\{f(X_\tau) - f(X_{\tau-0})\}^2)$$

for bounded f in $\underline{\underline{F}}_{(e)}$ and for f' a normalized contraction of f. Fix $\varepsilon > 0$ and define

$$\tau = \inf\{t > 0 : |f(X_t) - f(X_{t-0})| > \varepsilon\}.$$

Also let $\{M^\varepsilon f(t)\}$ and $\{M^\varepsilon f'(t)\}$ be the orthogonal complements in the sense of [35], VIII.3] to the contributions to $\{Mf(t)\}$ and $\{Mf'(t)\}$ from discontinuities at time σ when $|f(X_\sigma) - f(X_{\sigma-0})| > \varepsilon$. Then (11.24) can be written

(11.24') $$\mathscr{E}_x\{M^\varepsilon f'(\tau) - M^\varepsilon f'(0)\}^2$$

$$\leq \mathscr{E}_x\{M^\varepsilon f(\tau) - M^\varepsilon f(0)\}^2$$

and it follows from the approximation result [36, p. 91] that

(11.25) $$\langle M^\varepsilon f'\rangle(dt) \leq \langle M^\varepsilon f\rangle(dt)$$

on the random interval $[0,\tau]$. The restriction to this interval can be removed by an obvious "piecing together argument." Finally a passage to the limit $\varepsilon \downarrow 0$ and then a passage to the limit in f and f' establishes

<u>Theorem 11.7.</u> <u>If</u> f <u>is in</u> $\underline{\underline{F}}_{(e)}$ <u>and if</u> f' <u>is a normalized contraction of</u> f <u>then</u>

(11.26) $$\langle M_c f'\rangle(dt) \leq \langle M_c f\rangle(dt). ///$$

<u>11.2. Definition.</u> The <u>diffusion form</u> is defined on $\underline{\underline{F}}_{(e)}$ by

$$D(f,f) = E(f,f) - \frac{1}{2}\int\int J(dy,dz)\,\{f(y)-f(z)\}^2$$
$$- \int \kappa(dy)\, f^2(y). \quad ///$$

In the transient case clearly

(11.27) $$D(f,f) = \frac{1}{2}\,\mathcal{E}\{M_c f(\zeta) - M_c f(\zeta^*)\}^2$$

and in the recurrent case for $u > 0$

(11.27') $$D(f,f) = \frac{1}{2}\,u\int dx\ \mathcal{E}_x\{M_c f(R_u) - M_c f(0)\}^2.$$

Thus Theorem 11.7 has the

<u>Corollary 11.8.</u> <u>If</u> f <u>is in</u> $\underline{\underline{F}}_{(e)}$ <u>and if</u> f' <u>is a normalized contraction of</u> f <u>then</u>

(11.28) $$D(f',f') \le D(f,f). \quad ///$$

With the help of Theorem 9.3 it is easy to check that for each f in $\underline{\underline{F}}_{(e)}$ there exists a unique measure $\langle A_c f\rangle(dx)$ such that

(11.29) $$\langle M_c f\rangle(dt) = a(\langle A_c f\rangle;\ dt).$$

Of course

(11.30) $$\langle A_c f'\rangle\,(dt) \le \langle A_c f\rangle\,(dx)$$

if f' is a normalized contraction of f. Also it follows from the results in section 10 that

(11.31) $$\langle Af\rangle(dx) = \langle A_c f\rangle(dx) + \int J(dx,dy)\ \{f(y)-f(x)\}^2.$$

We finish by establishing the basic connection between "locality" for Dirichlet forms and continuous sample paths for the corresponding processes.

See [25] for an entirely different approach.

11.3. Definition. Two functions f, f' are strongly separated if there exists a metric d for $\underline{X}$ and disjoint Borel sets A, A' such that

$$f = 0 \quad \text{almost everywhere on } \underline{X} - A$$

$$f' = 0 \quad \text{almost everywhere on } \underline{X} - A'$$

$$d(A, A') = \inf \{d(x, x') \; ; \; x \in A, \; x' \in A'\} > 0. \; ///$$

Theorem 11.9. If f, f' in $\underline{F}_{(e)}$ are strongly separated then

(11.32) $$\langle M_c f, M_c f' \rangle (t) = 0$$

(11.33) $$D(f, f') = 0. \; ///$$

Remark. Of course $\langle M_c f, M_c f' \rangle (t)$ has the same relation to the product $M_c f(t) \, M_c f'(t)$ that $\langle M_c f \rangle (t)$ has to the square $\{M_c f(t)\}^2$. It can be obtained explicitly from

$$\langle M_c f, M_c f' \rangle (t) = (1/4) \langle M_c (f+f') \rangle (t) - (1/4) \langle M_c (f-f') \rangle (t). \; ///$$

Proof. It suffices to establish (11.32) since (11.33) then follows upon integrating. Let A, A' and d be as in Definition 11.3. Choose $\varepsilon > 0$ such that $d(A, A') > 2\varepsilon$ and let τ be any stopping time such that $d(X_0, X_t) \le \varepsilon$ for $0 \le t < \tau$. It follows from (11.23) that for quasi-every x

$$\mathcal{E}_x \{Mf(\tau) - Mf(0)\} \{Mf'(\tau) - Mf'(0)\}$$
$$= \mathcal{E}_x \{f(X_\tau) - f(X_{\tau-0})\} \{f'(X_\tau) - f'(X_{\tau-0})\} .$$

But obviously

$$\mathcal{E}_x \Sigma_{0 < t < \tau} \{f(X_t) - f(X_{t-0})\} \{f'(X_t) - f'(X_{t-0})\} = 0$$

and so 11.13

$$\mathcal{E}_x\{M_c f(\tau) - M_c f(0)\}\{M_c f'(\tau) - M_c f'(0)\} = 0$$

or equivalently

$$\mathcal{E}_x \langle M_c f, M_c f' \rangle (\tau) = 0.$$

The expectation $\mathcal{E}_x$ can be "stripped off" as in the proof of Theorem 11.6 and again the theorem follows by a "piecing together" argument. ///

11.4. Definition. The Dirichlet space $(\underline{\underline{F}}, E)$ is local if $E(f, f') = 0$ whenever f, f' in $\underline{\underline{F}} \cap C_{com}(\underline{\underline{X}})$ have disjoint supports. ///

Theorem 11.10. The following are equivalent for a regular Dirichlet space $(\underline{\underline{F}}, E)$.

(i) $(\underline{\underline{F}}, E)$ is local.

(ii) $J = 0$.

(iii) The trajectory X_t is continuous for $0 \leq t < \zeta$. ///

Proof. Equivalence of (ii) and (iii) is clear from the very definition of J. If (iii) is true then $H^M f = f$ whenever M contains the support of f and (i) follows since H^M is an orthogonal projector relative to E. It only remains to prove that (i) implies (ii). But if f, f' in $\underline{\underline{F}} \cap C_{com}(\underline{\underline{X}})$ have disjoint supports then certainly they are strongly separated and so by Theorem 11.9

$$E(f, f') = \frac{1}{2} \int\int J(dy, dz)\{f(y) - f(z)\}\{f'(y) - f'(z)\} \tag{11.34}$$

$$= - \int\int J(dy, dz)\, f(y) f'(z).$$

11.14

If $J \neq 0$ it is easy to use regularity of $(\underline{\underline{F}}, E)$ to construct f, f' so that the right side of (11.34) is nonzero and the theorem is proved. ///

12. Characterization of $\varkappa$ and J.

Theorem 12.1. *Let $\lambda(dx)$ be a Radon measure charging no polar set such that*

$$(12.1)\qquad E(f',f') - \int\lambda(dx)\, f'^2(x) \le E(f,f) - \int\lambda(dx)\, f^2(x)$$

whenever f is in $\underline{\underline{F}}_{(e)}$ and f' is a normalized contraction of f. Then $\lambda(dx) \le \varkappa(dx)$. ///

Theorem 12.2. *Let $L(dy,dz)$ be a symmetric Borel measure on $\underline{\underline{X}} \times \underline{\underline{X}}$ satisfying*

$$(12.2)\qquad L(\{(x,x),\ x \in \underline{\underline{X}}\}) = 0$$

$$(12.3)\qquad L(\Gamma \times \underline{\underline{X}}) = 0$$

whenever Γ is a polar subset of $\underline{\underline{X}}$. Assume that

$$(12.4)\qquad E(f',f') - \frac{1}{2}\int\int L(dy,dz)\,\{f'(y)-f'(z)\}^2$$

$$\le E(f,f) - \frac{1}{2}\int\int L(dy,dz)\,\{f(y)-f(z)\}^2$$

whenever f is in $\underline{\underline{F}}_{(e)}$ and f' is a normalized contraction of f. Then $L(dy,dz) \le J(dy,dz)$. ///

Theorem 12.1 characterizes the killing measure $\varkappa(dx)$ as the maximal measure charging no polar set and satisfying (12.1). Theorem 12.2 characterizes the Lévy measure $J(dy,dz)$ as the maximal symmetric measure on $\underline{\underline{X}} \times \underline{\underline{X}}$ satisfying (12.2), (12.3) and (12.4). Condition (12.3) is needed since (12.4) does not control any mass of $L(dy,dz)$ concentrated on the diagonal.

12.2

In proving Theorem 12.1 we first use our results on random time change to reduce to the case where λ is absolutely continuous and then it is easy to reduce to the case $\lambda(dx) = \lambda(x)dx$ with λ a bounded function. Put

$$E^*(f,f) = E(f,f) - \int dx\, \lambda(x) f^2(x)$$

for f in $\underline{\underline{F}}_{(e)}$ and note that E_1^* is equivalent to E_1 and that (F, E^*) is a regular Dirichlet space. For φ, ψ in $L^2(dx)$ and for $u > 0$

$$(12.5) \qquad \int dx\, G_u^* \varphi(x) \psi(x)$$

$$= E_u(G_u^* \varphi, G_u \psi)$$

$$= E_u^* (G_u^* \varphi, G_u \psi) + \int dx\, \lambda(x)\, G_u^* \varphi(x)\, G_u \psi(x)$$

$$= \int dx\, \{ G_u \varphi(x) + G_u \lambda\, G_u^* \varphi(x) \} \psi(x) .$$

Therefore

$$G_u^* = G_u + G_u\, \lambda\, G_u^*$$

and it follows that

$$P_t f(x) = \mathcal{E}_x^* \exp\{ - \int_0^t ds\, \lambda(X_s) \} f(X_t)$$

and by an obvious probabilistic argument $\varkappa(dx)$ dominates $\lambda(x)\,dx$. ///

The same basic idea suffices to prove Theorem 12.2. If f, f' in $\underline{\underline{F}}_{(e)}$ have disjoint supports then

$$\frac{1}{2} \int L(dy,dz)\, \{f(y) - f(z)\} \{f'(y) - f'(z)\}$$

$$= - \iint L(dy,dz)\, f(y) f'(z)$$

and it follows from regularity of $\underline{\underline{F}}$ that $L(D \times D') < +\infty$ whenever D, D' are open with disjoint compact closures. We can replace L by its restriction to $(D \times D') \cup (D' \times D)$ and so there is no loss of generality in assuming that L itself is totally bounded. Then L has a unique representation

$$L(dy, dz) = \ell(dy)\, L(y, dz)$$

with $\ell(dy)$ a bounded Radon measure charging no polar set and $L(y, \cdot)$ a family of probabilities satisfying the usual regularity conditions. (See the paragraph following (10.10).) Again we use random time change to reduce to the special case $\ell(dy) = \ell(y)dy$ with ℓ bounded. For f in $\underline{\underline{F}}$ define

$$E^*(f,f) = E(f,f) - \frac{1}{2} \iint L(dy,dz)\, \{f(y) - f(z)\}^2$$

$$E^\sim(f,f) = E^*(f,f) + \int dy\, \ell(y) f^2(y)$$

$$= E(f,f) + \int dy\, \ell(y) \int L(y,dz)\, f(y)f(z) .$$

Again E_1^* and $E_1^\sim$ are equivalent to E_1 and $(\underline{\underline{F}}, E^*), (\underline{\underline{F}}, E^\sim)$ are regular Dirichlet spaces. A computation analagous to (12.5) establishes

$$G_u f = G_u^\sim f + G_u^\sim \ell \int L(\cdot, dz)\, G_u f(z)$$

and therefore

$$P_t f(x) = \mathcal{E}_x^\sim f(X_t) + \mathcal{E}_x^\sim \int_0^t ds\, \ell(X_s) \int L(X_s, dy)\, P_{t-s} f(y).$$

The proof of Theorem 12.1 shows that

$$P_t^\sim f(x) = \mathcal{E}_x^* \exp\{-\int_0^t \ell(X_s)\}\, f(X_t)$$

and therefore also

(12.6)
$$P_t f(x) = \mathcal{E}_x^* \exp\{-\int_0^t \ell(X_s)\} f(X_t)$$

$$+ \mathcal{E}_x^* \int_0^t ds \ \exp\{-\int_0^s du\ \ell(X_u)\}\ \ell(X_s) \int L(X_s, dy)\ P_{t-s} f(y).$$

But then for f bounded $P_t f$ must be the unique bounded solution of (12.6) and it follows by an obvious sample space construction that $J(dy, dz)$ dominates $dy\ \ell(y)\ L(y, dz)$. ///

Remark 1. The appeal to results on random time change can be avoided at the cost of refining the arguments in this section. ///

Remark 2. Theorem 12.2 can be used to obtain the main results in [45] without the side conditions imposed there. ///

Chapter III. Structure Theory

Some of the results were first established in [44], [46] and [47] using quite different techniques.

The results from Chapter III enable us to extend the Dirichlet form E to functions which are locally in $\underline{\underline{F}}$. The reflected Dirichlet space $\underline{\underline{F}}^{ref}$ is the set of such functions for which the extension is finite. In Section 13 we establish a "generalized Green's formula" which is used to analyze $\underline{\underline{F}}^{ref}$ in Section 14.

In Section 15 we define the local generator $\mathfrak{a}$ and prove the "First Structure Theorem." This states that the reflected space $\underline{\underline{F}}^{ref}$ serves as a kind of "classifying space" for Dirichlet spaces $(\underline{\underline{F}}^{\sim}, \tilde{E})$ with generator $A^{\sim}$ contained in $\mathfrak{a}$. The basic tool is an estimate which is established in exactly the same way as in [20]. It is the results in Chapter II which permit this direct extension of Fukushima's technique. An entirely different approach was used in [46]. The needed estimate was established first in a local context by analyzing certain time changed processes and then globally by passage to the limit.

The recurrent case is treated separately in Section 16. The main result is that $\underline{\underline{F}}^{ref} = \underline{\underline{F}}_{(e)}$ and the First Structure Theorem "collapses."

Section 17 is intended to motivate our continuing on the to Second and Third Structure Theorems. A semigroup $P_t^{\sim}$ is said to be an extension of P_t if P_t can be recovered by absorbtion upon exiting from an appropriate subset. The point is that an extension is covered by the First Structure Theorem if and only if "something new happens" only after a "drift to the dead point at infinite." Suppressing "jumps to the dead point" or replacing them with "jumps to someplace else" is not permitted. This is illustrated in the context of Markov chains. We consider a given symmetric Markov chain with stable states as an extension of the associated minimal one and we show that it is covered by the first structure theorem if and only if it satisfies the appropriate Kolmogorov equation.

Sections 18 and 19 are of a preparatory nature. In Section 18 we introduce the enveloping Dirichlet space $\underline{\underline{F}}^{env}$. This is derived from the given Dirichlet space by combining "resurrection" with "reflection." It conincides with the reflected space $\underline{\underline{F}}^{ref}$ only when the killing measure $\kappa = 0$. In Section 19 we introduce Fukushima's concept [21] of equivalent regular representations and the associated quasi-homeomorphism.

Section 20 contains the Second Structure Theorem which is the main result in this volume. It completes the First Structure Theorem and extends it scope considerably. A symmetric semigroup $P_t^{\sim}$ (Dirichlet space $(\underline{\underline{F}}^{\sim}, E^{\sim})$) is an extension of P_t(of $(\underline{\underline{F}},E)$) if P_t (if $(\underline{\underline{F}},E)$) can be recovered by absorption upon exiting from an appropriate subset. A semigroup $P_t^{\sim}$ is an extension if and only if $\underline{\underline{F}}_b$ is an ideal in the associated $\underline{\underline{F}}_b^{\sim}$ and $E^{\sim} = E$ on $\underline{\underline{F}}$. Every extension $(\underline{\underline{F}}^{\sim}, E^{\sim})$ is classified by at least one modified reflected space $(\underline{\underline{F}}^{ref\sim}, E^{ref\sim})$ in which case we say that $(\underline{\underline{F}}^{\sim}, E^{\sim})$ is subordinate to $(\underline{\underline{F}}^{ref\sim}, E^{ref\sim})$. These modified spaces can be defined directly as is the reflected space in Section 14 once we specify a pair (Δ, μ) satisfying a particular condition where Δ is a boundary for $\underline{\underline{X}}$ and where $\mu(x,d\cdot)$ measures the intensity for jumping to Δ instead of to the dead point. The extensions which are subordinate to a given modified reflected space are classified by certain Dirichlet spaces which live on Δ . For the special case when the modified reflected space is $(\underline{\underline{F}}_a^{ref}, E)$ the Second Structure Theorem implies the First Structure Theorem together with a converse and a classification scheme.

The Third Structure Theorem in Section 21 is applicable to every symmetric submarkovian semigroup which dominates the given one. This turns out to be a rather minor extension of the Second Structure Theorem. The most general such semigroup is obtained by first "suppressing jumps to the dead point" and/or replacing them by "jumps to somewhere else" in the given state space and then taking an extension.

13. Preliminary Formula

In Sections 13 through 15 we assume that $(\underline{\underline{F}}, E)$ is transient. The recurrent case will be treated separately in Section 16.

In this section we derive a formula which will play a fundamental role in the remainder of the volume Throughout the section D is an open subset of $\underline{\underline{X}}$ such that $M = \underline{\underline{X}} - D$ is nonpolar. Also we restrict attention to functions f having a representation

$$(13.1) \qquad f = f_0 + h$$

where f_0 belongs to the extended absorbed Dirichlet space $\underline{\underline{F}}^D_{(e)}$ and where h satisfies

$$(13.2) \qquad h(x) = \mathcal{E}_x [\sigma(M) < +\infty;\ h(X_{\sigma(M)})] + \mathcal{E}_x[\sigma(M) = +\infty;\ X_{\zeta-0} = \partial;\ \Phi]$$

for quasi-every x in D. Here h is specified and finite up to quasi-equivalence on D and up to $\Pi_1^M (1_D \cdot dx)$ equivalence on M and Φ is a random variable on the standard sample space Ω which is nonvanishing only on the set where $\sigma^*(M) < \zeta$, and $X_{\zeta-0} = \partial$; also

$$(13.3) \qquad \theta_t \Phi = \Phi \qquad \text{on } [t < \zeta].$$

Of course it follows from (13.3) that Φ is well defined on the extended sample space Ω_∞. We also impose the auxiliary condition

$$(13.4) \qquad \mathcal{E}_x[\sigma(M) < +\infty;\ h^2(X_{\sigma(M)})] + \mathcal{E}_x[\sigma(M) = +\infty;\ X_{\zeta-0} = \partial;\ \Phi^2] < +\infty$$

for quasi-every x in D.

In the next few sections we will need only the special case when D has compact closure and necessarily $\Phi = 0$. However the general version will be used later. Notice that by Theorem 7.3 every f in $\underline{\underline{F}}_{(e)}$ has a representation (13.1) with $\Phi = 0$, even if D does not have compact closure.

The random set

$$\{t > \sigma(M) : X_t \varepsilon D \text{ and } X_{t-0} \varepsilon D\}$$

is a finite or countable union of disjoint intervals $I = (e, r)$. Index the intervals $I_i = (e(i), r(i))$ so that the entrance times $e(i)$ and the return times $r(i)$ are Borel measurable. In general this indexing does not respect the "natural ordering" and neither the $e(i)$ nor the $r(i)$ are stopping times. (See however Sections 1 and 9 in [47] and Sections 3 and 5 in [48] where the natural order is respected and where the entrance and return times are stopping times.) We introduce the special notation

(13.5) $$h_{ref}(X_{\sigma(M)}) = I[\sigma(M) < +\infty] \; h(X_{\sigma(M)}) + \Phi$$

and for each i

(13.5') $$h_{ref}(X_{r(i)}) = I[r(i) < \zeta] \, h(X_{r(i)}) + I(r(i) = \zeta) \, \Phi$$

and then

(13.6) $$Mf(t) = Mf_o(t) + h(X_t) \quad \text{for} \quad X_t \varepsilon D$$

$$Mf(\sigma(M)) = Mf_o(\sigma(M)) + h_{ref}(X_{\sigma(M)})$$

$$Mf(r(i)) = Mf_o(r(i)) + h_{ref}(X_{r(i)}).$$

These functionals are martingales "along excursions into D" and so because of (13.4) the corresponding random measures $\langle Mf \rangle (dt)$ are well defined for the set of t such that $X_t \varepsilon D$. (This is true both on Ω and on Ω_∞.)

<u>Theorem 13.1.</u> <u>Assume that</u> $(\underline{F}, E)$ <u>is transient and let</u> f <u>have the representation</u> (13.1) <u>with</u> h <u>satisfying the auxiliary condition</u> (13.4). <u>Then</u>

$$(13.7)\quad \tfrac{1}{2}\,\mathscr{E}\int_{\zeta^*}^{\zeta} \langle Mf\rangle (dt)\, 1_D(X_t) + \tfrac{1}{2}\,\mathscr{E}\, I(X_{\zeta^*}\in D)\, f^2(X_{\zeta^*})$$

$$+ \tfrac{1}{2}\,\mathscr{E}\,\Sigma_i\{f(X_{e(i)}) - f(X_{e(i)-0})\}^2$$

$$= E(f_0, f_0) + \tfrac{1}{2}\,\mathscr{E}\,\Sigma_i\{h_{ref}(X_{r(i)}) - h(X_{e(i)-0})\}^2$$

$$+ \tfrac{1}{2}\,\mathscr{E}\, I(X_{\zeta^*}\in D)\, h^2_{ref}(X_{\sigma(M)})$$

$$+ \tfrac{1}{2}\,\mathscr{E}\, I(X_{\zeta^*} = \partial\,;\ \sigma(M) > \zeta^*)\,\{h_{ref}(X_{\sigma(M)}) - \Phi\cdot\rho\}^2.$$

Equivalently

$$(13.7')\quad \tfrac{1}{2}\,\mathscr{E}\int_{\zeta^*}^{\zeta} \langle M_c f\rangle (dt)\, 1_D(X_t) + \int_D \varkappa(dy)\, f^2(y)$$

$$+ \iint_{D\times M} J(dy, dz)\,\{f(y) - f(z)\}^2$$

$$+ \tfrac{1}{2}\iint_{D\times D} J(dy, dz)\,\{f(y) - f(z)\}^2$$

$$= E(f_0, f_0) + \tfrac{1}{2}\,\mathscr{E}\,\Sigma_i\,\{h_{ref}(X_{r(i)}) - h(X_{e(i)-0})\}^2$$

$$+ \tfrac{1}{2}\,\mathscr{E}\, I(X_{\zeta^*}\in D)\, h^2_{ref}(X_{\sigma(M)})$$

$$+ \tfrac{1}{2}\,\mathscr{E}\, I(X_{\zeta^*} = \partial\,;\ \sigma(M) > \zeta^*)\}\, h_{ref}(X_{\sigma(M)}) - \Phi\cdot\rho\}^2.\quad ///$$

Proof. It suffices to consider the special case when h is bounded and when $f_0 = G^D\varphi$ with φ bounded and integrable and with f_0 bounded. Then

(13.8) $$\frac{1}{2}\,\mathcal{E}_x\int_0^{\sigma(M)} \langle Mf\rangle\,(dt)$$

$$= \frac{1}{2}\,\mathcal{E}_x\{Mf(\sigma(M)) - Mf(0)\}^2$$

$$= \frac{1}{2}\,\mathcal{E}_x\{h_{ref}(X_{\sigma(M)}) - f(X_0) + \int_0^{\sigma(M)} dt\,\varphi(X_t)\}^2$$

$$= \mathcal{E}_x\int_0^{\sigma(M)} dt\;\varphi(X_t)\;G^D\varphi(X_t)$$

$$+ \frac{1}{2}\,\mathcal{E}_x\;\{h_{ref}(X_{\sigma(M)}) - f(X_0)\}^2$$

$$+ \mathcal{E}_x\{h_{ref}(X_{\sigma(M)}) - f(X_0)\}\int_0^{\sigma(M)} dt\;\varphi(X_t)\,.$$

We will also need a modified form of (13.8). The right side of (13.8)

$$= \mathcal{E}_x\int_0^{\sigma(M)} dt\,\varphi(X_t)\,G^D\varphi(X_t)$$

$$+\frac{1}{2}\,\mathcal{E}_x\,h^2_{ref}(X_{\sigma(M)}) - \frac{1}{2}\,\mathcal{E}_x\,f^2(X_0)$$

$$- \mathcal{E}_x f(X_0)\{h_{ref}(X_{\sigma(M)}) - f(X_0)\}$$

$$+ \mathcal{E}_x\,h_{ref}(X_{\sigma(M)})\int_0^{\sigma(M)} dt\,\varphi(X_t)$$

$$- \mathcal{E}_x f(X_0)\int_0^{\sigma(M)} dt\,\varphi(X_t)$$

and since

$$\mathcal{E}_x f(X_0)\{h_{ref}(X_{\sigma(M)}) + \int_0^{\sigma(M)} dt\,\varphi(X_t) - f(X_0)\} = 0$$

there follows

$$(13.9)\qquad \frac{1}{2}\,\mathcal{E}_x \int_0^{\sigma(M)} \langle Mf\rangle(dt) + \frac{1}{2}\,\mathcal{E}_x f^2(X_0)$$

$$= \mathcal{E}_x \int_0^{\sigma(M)} dt\, \varphi(X_t)\, G^D \varphi(X_t)$$

$$+ \frac{1}{2}\,\mathcal{E}_x h^2_{\mathrm{ref}}(X_{\sigma(M)})$$

$$+ \mathcal{E}_x h_{\mathrm{ref}}(X_{\sigma(M)}) \int_0^{\sigma(M)} dt\, \varphi(X_t).$$

Next fix D' open with compact closure contained in D and define

$$(13.10)\qquad \varepsilon(1) = \inf\{t > \sigma(M) : X_t \varepsilon D'\}$$

$$\rho(1) = \inf\{t > \varepsilon(1) : X_t \varepsilon M\}$$

$$\varepsilon(2) = \inf\{t > \rho(1) : X_t \varepsilon D'\}$$

etc.

with the usual understanding that these times are $+\infty$ when not otherwise defined. A computation analagous to (13.8) establishes for each i

$$(13.11)\qquad \frac{1}{2}\,\mathcal{E}_x \int_{\varepsilon(i)}^{\rho(i)} \langle Mf\rangle(dt)$$

$$= \mathcal{E}_x \int_{\varepsilon(i)}^{\rho(i)} dt\, \varphi(X_t)\, G^D \varphi(X_t)$$

$$+ \frac{1}{2}\,\mathcal{E}_x \{h_{\mathrm{ref}}(X_{\rho(i)}) - f(X_{\varepsilon(i)})\}^2$$

$$+ \mathcal{E}_x \{h_{\mathrm{ref}}(X_{\rho(i)}) - f(X_{\varepsilon(i)})\} \int_{\varepsilon(i)}^{\rho(i)} dt\, \varphi(X_t)$$

with the obvious understanding when $\varepsilon(i) = +\infty$. To transform this we assume temporarilly that the random intervals $(e(i), r(i))$ are labeled so that

$$(13.10')\qquad e(i) = \sup\{t < \varepsilon(i) : X_t \text{ or } X_{t-0}\ \varepsilon\ M\}$$

whenever $\varepsilon(i) < +\infty$. For each such i

$$\mathcal{E}_x\{f(X_{\epsilon(i)}) - f(X_{e(i)-0})\}\{h_{ref}(X_{r(i)}) - f(X_{\epsilon(i)}) + \int_{\epsilon(i)}^{r(i)} dt\, \varphi(X_t)\} = 0$$

and since $f(X_{e(i)-0}) = h(X_{e(i)-0})$ the same manipulations as above yield

$$(13.12)\quad \frac{1}{2}\mathcal{E}_x \int_{\epsilon(i)}^{\rho(i)} \langle Mf\rangle (dt) + \frac{1}{2}\mathcal{E}_x\{f(X_{\epsilon(i)}) - f(X_{e(i)-0})\}^2$$

$$= \mathcal{E}_x \int_{\epsilon(i)}^{\rho(i)} dt\, \varphi(X_t)\, G^D\varphi(X_t)$$

$$+ \frac{1}{2}\mathcal{E}_x\{h_{ref}(X_{r(i)}) - h(X_{e(i)-0})\}^2$$

$$+ \mathcal{E}_x\{h_{ref}(X_{r(i)}) - h(X_{e(i)-0})\} \int_{\epsilon(i)}^{r(i)} dt\, \varphi(X_t) .$$

Now that (13.12) is established we change our conventions on labeling. Let the labels $e(i)$, $r(i)$ be fixed independent of D' and then for each D' let

$$e'(i) = \inf\{t : e(i) < t < r(i) : X_t \in D'\}$$

with the understandin that $e'(i) = +\infty$ if not otherwise defined. (Note that the $e'(i)$ are in general not stopping times.) Summing (13.12) over i we get

$$(13.13)\quad \frac{1}{2}\mathcal{E}_x \Sigma_i' \int_{e'(i)}^{r(i)} \langle Mf\rangle (dt) + \frac{1}{2}\mathcal{E}_x \Sigma_i' \{f(X_{e'(i)}) - f(X_{e(i)-0})\}^2$$

$$= \mathcal{E}_x \Sigma_i' \int_{e'(i)}^{r(i)} dt\, \varphi(X_t)\, G^D\varphi(X_t)$$

$$+ \frac{1}{2}\mathcal{E}_x \Sigma_i' \{h_{ref}(X_{r(i)}) - h(X_{e(i)-0})\}^2$$

$$+ \mathcal{E}_x \Sigma_i' \{h_{ref}(X_{r(i)}) - h(X_{e(i)-0})\} \int_{e'(i)}^{r(i)} dt\, \varphi(X_t)$$

where for typographical convenience we have used Σ_i' to denote summation over i such that $e'(i) < +\infty$. If we knew that

$$(13.14) \quad \mathscr{E}_x \sup_{D'} \Sigma_i' \{f(X_{e'(i)}) - f(X_{e(i)-0})\}^2 < +\infty$$

then we could pass to the limit $D' \uparrow D$ and get

$$(13.15) \quad \tfrac{1}{2} \mathscr{E}_x \int_{\sigma(M)}^{\zeta} \langle Mf\rangle (dt)\, 1_D(X_t) + \tfrac{1}{2} \mathscr{E}_x \Sigma_i \{f(X_{e(i)}) - f(X_{e(i)-0})\}^2$$

$$= \mathscr{E}_x \int_{\sigma(M)}^{\zeta} dt\, \varphi(X_t)\, G^D \varphi(X_t)$$

$$+ \tfrac{1}{2} \mathscr{E}_x \Sigma_i \{ h_{ref}(X_{r(i)}) - h(X_{e(i)-0}) \}^2$$

$$+ \mathscr{E}_x \Sigma_i \{h_{ref}(X_{r(i)}) - h(X_{e(i)-0})\} \int_{e(i)}^{r(i)} dt\, \varphi(X_t).$$

Since

$$f(X_{e'(i)}) - f(X_{e(i)-0})$$

$$= \{f(X_{e(i)}) - f(X_{e(i)-0})\} + \{Mf(e'(i)) - Mf(e(i))\}$$

$$- \int_{e(i)}^{e'(i)} dt\, \varphi(X_t)$$

it would suffice to establish

$$(13.14a) \quad \mathscr{E}_x \Sigma_i \{f(X_{e(i)}) - f(X_{e(i)-0})\}^2 < +\infty$$

$$(13.14b) \quad \mathscr{E}_x \Sigma_i \{ \int_{e(i)}^{r(i)} dt\, \varphi(X_t)\}^2 < +\infty.$$

$$(13.14c) \quad \mathscr{E}_x \Sigma_i \sup_{e(i) < t \leq r(i)} \{Mf(t) - Mf(e(i))\}^2 < +\infty .$$

To establish (13.14a) note that by Theorem 6.4-(i) and by Meyer's results on square integrable martingales [33, VIII. 3]

$$(13.16)\qquad \mathcal{E}_x\, \Sigma_i \{ f(X_{e(i)}) - f(X_{e(i)-0}) \}^2$$

$$= \mathcal{E}_x\, \Sigma_i \{ Mf(e(i)) - Mf(e(i)-0) \}^2$$

$$\leq \mathcal{E}_x \{ Mf(\infty) - Mf(0) \}^2 .$$

$$= \mathcal{E}_x \{ \int_0^\zeta dt\ \varphi(X_t) + \Phi - f(X_0) \}^2$$

which is finite for quasi-every x by our assumptions on φ, $G^D\varphi$. For (13.14b) it suffices to observe that

$$\mathcal{E}_x \Sigma_i \{ \int_{e(i)}^{r(i)} dt\ \varphi(X_t) \}^2$$

$$\leq \mathcal{E}_x \{ \int_0^\zeta dt\ \varphi(X_t) \}^2$$

$$\leq 2\, \mathcal{E}_x \int_0^\zeta dt\ \varphi(X_t)\, G^D\varphi(X_t) \quad .$$

For (13.14c) we temporarilly return to our notations (13.10) and note that by Fatou's lemma it suffices to estimate

$$(13.17)\qquad \mathcal{E}_x \Sigma_i \sup_{\varepsilon(i) < t \leq r(i)} \{ Mf(t) - Mf(\varepsilon(i)) \}^2$$

independent of D'. But by the maximal inequality for square integrable martingales [35, p. 88], the expression (13.17) is dominated by

$$2\, \mathcal{E}_x\, \Sigma_i \{ Mf(r(i)) - Mf(\varepsilon(i)) \}^2$$

$$\leq 2\, \mathcal{E}_x \{ Mf(\infty) - Mf(0) \}^2 .$$

Thus (13.14c) and therefore (13.15) is established. Integrating (13.15) with respect to $L_k(dx)$ and passing to the limit $k \uparrow \infty$ we get

$$(13.18)\qquad \frac{1}{2}\,\mathcal{E}\int_{\sigma(M)}^{\zeta} \langle Mf\rangle\,(dt)\,1_D(X_t)$$

$$+\frac{1}{2}\,\mathcal{E}\,\Sigma_i\{f(X_{e(i)})-f(X_{e(i)-0})\}^2$$

$$=\mathcal{E}\int_{\sigma(M)}^{\zeta} dt\,\varphi(X_t)\,G^D\varphi(X_t)$$

$$+\frac{1}{2}\,\mathcal{E}\,\Sigma_i\{h_{ref}(X_{r(i)})-h(X_{e(i)-0})\}^2$$

$$+\mathcal{E}\,\Sigma_i\{h_{ref}(X_{r(i)})-h(X_{e(i)-0})\}\int_{e(i)}^{r(i)} dt\,\varphi(X_t).$$

For the contribution up to time $\sigma(M)$ we treat separately the sample paths in Ω_∞ where $[X_{\zeta^*}=\partial]$ and where $[X_{\zeta^*}\neq\partial]$. Integrating (13.9) over D with respect to $\varkappa(dx)$ we get

$$(13.19)\qquad \frac{1}{2}\,\mathcal{E}\,I(X_{\zeta^*}\in D)\int_0^{\sigma(M)} \langle Mf\rangle\,(dt)$$

$$+\frac{1}{2}\,\mathcal{E}\,I(X_{\zeta^*}\in D)\,f^2(X_{\zeta^*})$$

$$=\mathcal{E}\,I(X_{\zeta^*}\in D)\int_{\zeta^*}^{\sigma(M)} dt\,\varphi(X_t)\,G^D\varphi(X_t)$$

$$+\mathcal{E}\,I(X_{\zeta^*}\in D)\,h^2_{ref}(X_{\sigma(M)})$$

$$+\mathcal{E}\,I(X_{\zeta^*}\in D)h_{ref}(X_{\sigma(M)})\int_{\zeta^*}^{\sigma(M)} dt\,\varphi(X_t).$$

For each k let L_k^o be the unique measure in $\mathcal{M}$ such that

$$(13.20)\qquad H^k\{1-N\varkappa\} = NL_k^o\ .$$

Integrating (13.8) with respect to $L_k^o(dx)$ and passing to the limit $k\uparrow\infty$ we get

(13.21) $$\frac{1}{2}\mathscr{E}\, I(X_{\zeta *} = \partial;\ \sigma(M) > \zeta *) \int_{\zeta *}^{\sigma(M)} \langle Mf\rangle(dt)$$

$$= \mathscr{E}\, I(X_{\zeta *} = \partial\ ;\ \sigma(M) > \zeta *) \int_{\zeta *}^{\sigma(M)} dt\, \varphi(X_t)\, G^D\varphi(X_t)$$

$$+ \frac{1}{2}\mathscr{E}\, I(X_{\zeta *} = \partial)\, \{h_{ref}(X_{\sigma(M)}) - \Phi\cdot\rho\}^2$$

$$+ \mathscr{E}\, I(X_{\zeta *} = \partial\ ;\ \sigma(M) > \zeta *)\{h_{ref}(X_{\sigma(M)}) - \Phi\cdot\rho\} \int_{\zeta *}^{\sigma(M)} dt\, \varphi(X_t).$$

Finally (13.7) follows upon combining (13.18), (13.19) and (13.21) since

$$\mathscr{E}\, \Sigma_i\{h_{ref}(X_{r(i)}) - h(X_{e(i)-0})\} \int_{e(i)}^{r(i)} dt\, \varphi(X_t)$$

$$+ \mathscr{E}\, I(X_{\zeta *} \in D) h_{ref}(X_{\sigma(M)}) \int_{\zeta *}^{\sigma(M)} dt\, \varphi(X_t)$$

$$+ \mathscr{E}\, I(X_{\zeta *} = \partial\ ;\ \sigma(M) > \zeta *)\{h_{ref}(X_{\sigma(M)}) - \Phi\cdot\rho\} \int_{\zeta *}^{\sigma(M)} dt\, \varphi(X_t)$$

changes sign under time reversal and theorefore must vanish and since

$$\mathscr{E} \int_{\zeta *}^{\zeta} dt\, \varphi(X_t)\, G^D\varphi(X_t)$$

$$= \int d\mathbf{x}\, \varphi(\mathbf{x}) G^D\varphi(\mathbf{x})$$

$$= E(G^D\varphi, G^D\varphi). ///$$

14. The Reflected Dirichlet Space..

14.1. Definition. f belongs to the local Dirichlet space $\underline{\underline{F}}_{\ell oc}$ if for each x in X there exists a neighborhood U of x and a function f_U in $\underline{\underline{F}}$ such that $f = f_U$ quasi-everywhere on U. ///

It follows easily from (11.23) that for f in $\underline{\underline{F}}_{\ell oc}$ there exists a unique measure $\langle A_c f\rangle (dx)$ such that

$$\langle A_c f\rangle (dy) = \langle A_c f_U\rangle (dy) \qquad \text{on } U. \tag{14.1}$$

14.2. Definition. f in $\underline{\underline{F}}_{\ell oc}$ belongs to the reflected Dirichlet space $\underline{\underline{F}}^{ref}$ if

$$E(f,f) = \frac{1}{2}\int \langle A_c f\rangle (dy) + \tfrac{1}{2}\int\int J(dy,dz)\{f(y)-f(z)\}^2 + \int \varkappa(dy) f^2(y) \tag{14.2}$$

is finite. ///

Obviously $\underline{\underline{F}}_{(e)}$ is contained in $\underline{\underline{F}}_{ref}$ and (14.2) is consistent with previous definitions when f is in $\underline{\underline{F}}_{(e)}$. Thus (14.2) implicitly extends the Dirichlet form E from $\underline{\underline{F}}_{(e)}$ to $\underline{\underline{F}}^{ref}$.

Lemma 14.1. If f in $\underline{\underline{F}}_{\ell oc}$ is bounded and if D is open with compact closure, then there exists f_D in $\underline{\underline{F}}$ such that $f = f_D$ quasi-everyhere on D. ///

Proof. Obviously there exist open sets $U_1, \ldots, U_n$ and $U'_1, \ldots, U'_n$ and bounded functions $f_1, \ldots, f_n$ in $\underline{\underline{F}}$ such that each U_i is open with compact closure contained in U'_i and $f_i = f$ quasi-everywhere on U'_i and such that the union of the U_i cover D. For each i let ψ_i in $\underline{\underline{F}}$ satisfy $0 \leq \psi_i \leq 1$

and $\psi_i = 1$ on U_i and $\psi_i = 0$ on $\underline{\underline{X}} - U_i'$. Then $\psi = \Sigma_{i=1}^{n} \psi_i$ belongs to $\underline{\underline{F}}$ and so does $T\psi$ where T is the normalized contraction

(14.3) $$Tx = \begin{cases} x & \text{for } |x| \leq 1 \\ 1/x & \text{for } |x| > 1. \end{cases}$$

Also $\varphi_i = (T\psi)\,\psi_i$ belongs to $\underline{\underline{F}}$ and since $\Sigma_{i=1}^{n} \varphi_i = 1$ on D we can define

(14.4) $$f_D = \Sigma_{i=1}^{n} \varphi_i f_i. \quad ///$$

If f belongs to $\underline{\underline{F}}_{\ell oc}$ then the functional $M_c f(t)$ is well defined on Ω for $t < \zeta$ and also for $t = \zeta$ when $X_{\zeta - 0} \neq \partial$ by the condition

(14.5) $$M_c f(t) = M_c f_D(t) \qquad 0 \leq t \leq \sigma(\underline{\underline{X}}-D).$$

We also define

(14.6) $$\begin{aligned} Mf(t) = M_c f(t) &+ \Sigma_{0<r\leq t} \{f(X_r) - f(X_{r-0})\} \\ &- a(\textstyle\int J(d\cdot, dz)\{f(z) - f(\cdot)\}\,;\, t) \\ &+ a(f\cdot\kappa\,;\, t) \end{aligned}$$

when the right side makes sense-which is certainly the case for f in $\underline{\underline{F}}^{ref}$. Indeed for f in $\underline{\underline{F}}^{ref}$ it follows with the help of the martingale convergence theorem that the restriction on t can be removed and that these functionals are well defined on Ω_∞. The same is true for the corresponding functionals $\langle M_c f\rangle(t)$ and $\langle Mf\rangle(t)$. Also by regularity of $\underline{\underline{F}}$

(14.7) $$J(D, \underline{\underline{X}} - D') < +\infty$$

whenever D is open with closure compact and contained in the open set D' and therefore

$$(14.8) \qquad \int_D \langle A_c f\rangle (dx) + \tfrac{1}{2} \iint_{D\times D} J(dy,dz)\{f(y)-f(z)\}^2$$

$$+ \iint_{D\times M} J(dy,dz)\{f(y)-f(z)\}^2 + \int_D \kappa(dy) f^2(y)$$

is finite for bounded f in $\underline{\underline{F}}_{loc}$ and for open D having compact closure.

Lemma 14.2. Let f in $\underline{\underline{F}}_{loc}$ be bounded and let D be open with compact closure. Then $f-H^M f$ belongs to the absorbed space $\underline{\underline{F}}^D$ and

$$(14.9) \qquad E(f-H^M f, f-H^M f) \le \int_D \langle A_c f\rangle (dx) + \tfrac{1}{2} \iint_{D\times D} J(dy,dz)\{f(y)-f(z)\}^2$$

$$+ \iint_{D\times M} J(dy,dz)\{f(y)-f(z)\}^2 + \int_D \kappa(dy) f^2(y).$$

Moreover the restriction that f be bounded can be removed if it is known that the right side of (14.9) is finite. ///

Proof. Let D' be a second open set with compact closure such that D' contains the closure of D and let f' be as in Lemma 14.2 for D'. Obviously $f'-H^M f'$ belongs to the absorbed space $\underline{\underline{F}}^D_{(e)}$ and it follows in particular from Theorem 13.1 that

$$(14.10) \qquad E(f'-H^M f', f'-H^M f')$$

$$\le \tfrac{1}{2} \mathcal{E} \int_{\zeta^*}^{r} \langle Mf'\rangle (dt)\, 1_D(X_t)$$

$$+ \tfrac{1}{2} \mathcal{E}\, I(X_{\zeta^*} \in D)\, f'^2(X_{\zeta^*})$$

$$+ \tfrac{1}{2} \mathcal{E}\, \Sigma_i \{f'(X_{e(i)}) - f'(X_{e(i)-0})\}^2$$

$$= \tfrac{1}{2} \int_D \langle A_c f\rangle (dx) + \tfrac{1}{2} \iint_{D\times \underline{X}} J(dy,dz)\{f(y)-f'(z)\}^2$$

$$+\int_D \kappa(dy)\, f^2(y) + \tfrac{1}{2}\iint_{D\times M} J(dy,dz)\{f(y)-f'(z)\}^2$$

which is dominated by the right side of (14.9) together with $J(D,M')\,(\|f\|_\infty + \|f'\|_\infty)^2$. The first part of the lemma follows with the help of (14.7) after passage to the limit in D' since we can assume $\|f'\|_\infty \leq \|f\|_\infty$ and then $H^M f' \to H^M f$ quasi-everywhere on D. (See 1.6.1'.) The remainder of the lemma follows after an obvious passage to the limit in f. ///

In the course of proving Lemma 14.2 we established the following result which we state separately for convenient future reference.

Corollary 14.3. Let bounded f be in F^{loc} and let f_n, $n \geq 1$ be a sequence of uniformly bounded functions in F^{loc} such that $f_n \to f$ quasi-everywhere. Let D be open with compact closure $cl(D)$, let $M = \underline{X} - D$ and suppose there exists a neighborhood of $cl(D)$ on which every $f_n = f$ quasi-everywhere. Then the differences $\{f_n - H^M f_n\}$, $n \geq 1$ belong to the absorbed space $\underline{F}^{D(e)}$ and are bounded in E norm and therefore a subsequence converges to $f - H^M f$ relative to E.

14.3. Definition. A function h is harmonic on $\underline{X}$ if it is specified and finite up to quasi-equivalence and if $h = H^M h$ whenever $D = \underline{X} - M$ is open with compact closure in $\underline{X}$. ///

14.4. Definition. A terminal random variable is a Borel function Φ on Ω such that Φ is nonvanishing only on the set $[X_{\zeta-0} = \partial]$ and such that $\theta_t \Phi = \Phi$ on the set $[t < \zeta]$. ///

As in Section 13 we note that Φ is well defined on Ω_∞.

The basic stucture theorem for $\underline{F}^{ref}$ is

Theorem 14.4. Assume that $(\underline{F}, E)$ *is transient and let* f *belong to the reflected Dirichlet space* $\underline{F}^{ref}$. *Then* f *has a unique representation*

(14.11) $$f = f_o + h$$

with f_o *in* $\underline{F}_{(e)}$ *and* h *harmonic and*

(14.12) $$E(f, f) = E(f_o, f_o) + E(h, h).$$

Moreover

(14.13) $$h = \mathcal{E}\,\Phi$$

with Φ *a terminal variable and*

(14.4) $$E(h, h) = \tfrac{1}{2}\,\mathcal{E}\{\Phi - \Phi\cdot\rho\}^2. \quad ///$$

Proof. The existence of a decomposition (14.11) with $E(f_o, f_o) \leq E(f, f)$ follows immediately from Lemma 14.2 after passage to the limit in D. In proving (14.12) and (14.14) we adapt the technique of Section 13 and again it suffices to consider the special case when h is bounded and when $f_o = G\varphi$ with φ bounded and integrable and with f_o bounded. First

(14.15) $$\tfrac{1}{2}\,\mathcal{E}_x \langle Mf\rangle(\zeta)$$
$$= \tfrac{1}{2}\,\mathcal{E}_x\{\Phi - f(X_0) + \int_0^\zeta dt\, \varphi(X_t)\}^2$$
$$= \mathcal{E}_x \int_0^\zeta dt\, \varphi(X_t) G\varphi(X_t)$$
$$+ \tfrac{1}{2}\,\mathcal{E}_x\{\Phi - f(X_0)\}^2$$
$$+ \mathcal{E}_x\{\Phi - f(X_0)\} \int_0^\zeta dt\, \varphi(X_t).$$

Also

$$\tfrac{1}{2}\,\mathscr{E}_{\mathbf{x}} \langle Mf\rangle(\zeta)$$

$$= \mathscr{E}_{\mathbf{x}} \int_0^{\zeta} dt\, \varphi(X_t)\, G\, \varphi(X_t)$$

$$+ \tfrac{1}{2}\,\mathscr{E}_{\mathbf{x}}\, \Phi^2 - \mathscr{E}_{\mathbf{x}} f(X_0)\{\Phi - f(X_0)\}$$

$$- \tfrac{1}{2}\,\mathscr{E}_{\mathbf{x}} f^2(X_0) + \mathscr{E}_{\mathbf{x}}\, \Phi \int_0^{\zeta} dt\, \varphi(X_t)$$

$$- \mathscr{E}_{\mathbf{x}} f(X_0) \int_0^{\zeta} dt\, \varphi(X_t)$$

and since

$$-\mathscr{E}_{\mathbf{x}} f(X_0)\{\Phi - f(X_0)\} - \mathscr{E}_{\mathbf{x}} f(X_0) \int_0^{\zeta} dt\, \varphi(X_t) = 0$$

also

(14.15')
$$\tfrac{1}{2}\,\mathscr{E}_{\mathbf{x}} \langle Mf\rangle(\zeta)$$

$$= \mathscr{E}_{\mathbf{x}} \int_0^{\zeta} dt\, \varphi(X_t)\, G\varphi(X_t)$$

$$+ \tfrac{1}{2}\,\mathscr{E}_{\mathbf{x}}\, \Phi^2 - \tfrac{1}{2}\,\mathscr{E}_{\mathbf{x}}\, f^2(X_0)$$

$$+ \mathscr{E}_{\mathbf{x}}\, \Phi \int_0^{\zeta} dt \varphi(X_t).$$

Integrating (14.15') with respect $\varkappa(dx)$ we get

(14.16)
$$\tfrac{1}{2}\,\mathscr{E}\, I(X_{\zeta^*} \in \underline{X}) \int_{\zeta^*}^{\zeta} \langle Mf\rangle(dt)$$

$$= \mathscr{E}\, I(X_{\zeta^*} \in \underline{X}) \int_{\zeta^*}^{\zeta} dt\, \varphi(X_t) G\varphi(X_t)$$

$$+ \tfrac{1}{2}\,\mathscr{E}\, I(X_{\zeta^*} \in \underline{X})\, \Phi^2$$

$$- \tfrac{1}{2}\,\mathscr{E}\, I(X_{\zeta^*} \in \underline{X})\, \Phi \int_{\zeta^*}^{\zeta} dt\, \varphi(X_t).$$

Integrating (14.15) with respect to L_k^o (dx) (see (13.20)) and passing to the limit $k \uparrow \infty$ we get

$$(14.17) \quad \tfrac{1}{2}\mathcal{E}\, I(X_{\zeta *}=\partial) \int_{\zeta *}^{\zeta} \langle Mf\rangle\,(dt)$$

$$= \mathcal{E}\, I(X_{\zeta *} = \partial) \int_{\zeta *}^{\zeta} dt\, \varphi(X_t)\, G\varphi(X_t)$$

$$+ \tfrac{1}{2}\, \mathcal{E}\, I(X_{\zeta *} = \partial)\ \{\Phi - \Phi\cdot\rho\}^2$$

$$+ \mathcal{E}\, I(X_{\zeta *} = \partial)\{\Phi - \Phi\cdot\rho\} \int_{\zeta *}^{\zeta} dt\, \varphi(X_t).$$

The theorem follows upon combining (14.16) and (14.17) since

$$\mathcal{E}\, I(X_{\zeta *} \in \underline{X})\ \Phi \int_{\zeta *}^{\zeta} dt\, \varphi(X_t)$$

$$+ \mathcal{E}\, I(X_{\zeta *}=\partial)\ \{\Phi - \Phi\cdot\rho\} \int_{\zeta *}^{\zeta} dt\, \varphi(X_t)$$

changes sign under time reversal and therefore vanishes, since

$$\mathcal{E} \int_{\zeta *}^{\zeta} dt\, \varphi(X_t)\, G\varphi(X_t)$$

$$= \int dx\, \varphi(x)\, G\varphi(X)$$

$$= E(G\varphi, G\varphi),$$

since

$$\tfrac{1}{2}\, \mathcal{E}\, I(X_{\zeta *} \in \underline{X})\ \Phi^2 + \tfrac{1}{2}\, \mathcal{E}\, I(X_{\zeta *} = \partial)\{\Phi - \Phi\cdot\rho\}^2$$

$$= \tfrac{1}{2}\, \mathcal{E}\{\Phi - \Phi\cdot\rho\}^2,$$

and since

$$\tfrac{1}{2}\, \mathcal{E} \int_{\zeta *}^{\zeta} \langle Mf\rangle\,(dt) + \tfrac{1}{2}\, \mathcal{E}\, I(X_{\zeta *} \in \underline{X})\ f^2(X_{\zeta *})$$

$$= E(f, f). \quad ///$$

Remark. It follows in particular from (14.13) that $\underline{\underline{F}}_b$ is an ideal in $\underline{\underline{F}}_b^{ref}$. ///

14.5. Notation. For f in $\underline{\underline{F}}^{ref}$ let Hf be the unique harmonic function such that f-Hf is in $\underline{\underline{F}}_{(e)}$. Let γf be the unique terminal random variable such that $Hf = \mathcal{E}_{\cdot}\,\gamma f$. Also for $u > 0$ define

$$(14.18) \qquad H_u f(x) = \mathcal{E}_x\, e^{-u\zeta}\,\gamma f . \quad ///$$

14.6. Definition. The active reflected Dirichlet space is the intersection

$$\underline{\underline{F}}_a^{ref} = \underline{\underline{F}}^{ref} \cap L^2(dx). \quad ///$$

Obviously the active reflected space $\underline{\underline{F}}_a^{ref}$ is the reflected space corresponding to $(\underline{\underline{F}}, E_u)$ for $u > 0$. Thus if f belongs to $\underline{\underline{F}}_a^{ref}$ then by Theorem 14.4 f has a unique representation $f = f_u + h_u$ with f_u in $\underline{\underline{F}}$ and with h_u u-harmonic and $E_u(f, f) = E_u(f_u, f_u) + E_u(h_u, h_u)$. Also by Theorem 7.3-(i) and the martingale convergence theorem

$$(14.19) \qquad h_u(x) = \mathcal{E}_x\, I(X_{\zeta -0} = \partial)\, \mathrm{Lim}_{k\uparrow\infty}\, e^{-u\sigma(M_k)} f(X_{\sigma(M_k)})$$

for some choice of D_k as in paragraph 5.1. But also $f = f_o + Hf$ and from (14.19) it follows that actually $h_u = H_u f$. In particular $Hf - H_u f$ belongs to $\underline{\underline{F}}_{(e)}$ and is bounded in E norm independent of $u > 0$. But then

$$\mathrm{Lim}_{u\downarrow 0}\{Hf - H_u f\} = \mathcal{E}_{\cdot}\, I(\zeta = +\infty)\,\gamma f$$

must belong to $\underline{\underline{F}}_{(e)}$ and it follows that γf vanishes on the set $[\zeta = +\infty]$. We summarize in

Theorem 14.5. Assume that $(\underline{F}, E)$ is transient and let f belong to the active reflected Dirichlet space $\underline{F}_a^{ref}$.

(i) The terminal variable γf vanishes on the set $[\zeta = +\infty]$.

(ii) For $u > 0$ the function $f-H_u f$ belongs to $\underline{F}$ and

(14.20) $$E_u(f, f) = E_u(f-H_u f, f-H_u f) + E_u(H_u f, H_u f).$$

Thus H_u is the E_u orthogonal projector of $\underline{F}_a^{ref}$ onto the complement of $\underline{F}$. ///

15. First Structure Theorem.

As in Section 1 we denote by A the $L^2(dx)$ generator of the semigroup $P_t, t > 0$.

15.1. Definition. f belongs to the domain of the local generator $\mathcal{Q}$ if f has a representation

(15.1) $$f = f_o + h$$

with f_o in domain A and with h harmonic. In this case

(15.2) $$\mathcal{Q} f = A f_o \quad ///$$

It follows from the results in Section 14 that the representation (15.1) is unique when it exists and therefore (15.2) is unambiguous for f in domain $\mathcal{Q}$.

We begin with a preliminary lemma.

Lemma 15.1. If g is bounded and has compact support then

$$\mathrm{Lim}_{v \uparrow \infty} \int dx\, g(x) \mathcal{E}_x I(X_{\zeta -0} = \partial)\, v e^{-v\zeta} = 0. \quad ///$$

Proof. By (5.8)

$$\int dx\, g(x) \mathcal{E}_x I(X_{\zeta -0} = \partial)\, v e^{-v\zeta}$$

$$= \mathcal{E} I(\zeta < +\infty;\ X_{\zeta -0} = \partial) \int_{\zeta *}^{\zeta} dt\, g(X_t)\, v e^{-v(\zeta - t)}.$$

As $v \uparrow \infty$ the integral $\int_{\zeta *}^{\zeta} dt\, g(X_t)\, v e^{-v(\zeta - t)}$ converges to $\mathrm{Lim}_{t \uparrow \zeta} g(X_t)$ which is 0 on the set $[X_{\zeta -0} = \partial]$ and the lemma follows with the help of the dominated convergence theorem since $\int_{\zeta *}^{\zeta} dt\, g(X_t)\, v e^{-v\zeta} \le \|g\|_\infty$ and since the support of g has finite capacity. ///

Consider now a strongly continuous symmetric submarkovian semigroup $P_t^{\sim}, t > 0$ on $L^2(dx)$ such that the generator $A^{\sim}$ is contained in the local generator $\mathcal{O}$; that is, every f in domain $A^{\sim}$ is contained in domain $\mathcal{O}$ and $A^{\sim} f = \mathcal{O} f$. Let $G_u^{\sim}, u>0$ be the associated resolvent and let $(\underline{F}^{\sim}, E^{\sim})$ be the associated Dirichlet space.

Fix $u > 0$, consider bounded φ in $L^2(dx)$ and let $f = G_u^{\sim} \varphi - G_u \varphi$. Then f is in domain $\mathcal{O}$ and $\mathcal{O} f = uf$. With h, f_o as in (15.1) clearly $(u-A)f_o = uf_o - uf = -uh$; thus $f_o = -uG_u h$ and so $f = h - uG_u h$ is u-harmonic. It follows with the help of the martingale convergence theorem that actually $f = H_u G_u \varphi$ and so

$$(15.3) \qquad G_u^{\sim} \varphi = G_u \varphi + H_u G_u^{\sim} \varphi .$$

In particular $G_u^{\sim}$ dominates G_u and so by Lemma 1.1 $\underline{F}^{\sim}$ contains $\underline{F}$. Actually much more can be deduced from (15.3) with the help of Lemma 15.1. For f, g in $L^2(dx)$ with f bounded

$$u \int dx \; \{f(x) - uG_u f(x)\} \; g(x) - u \int dx \{f(x) - uG_u^{\sim} f(x)\} \, g(x)$$

$$= u^2 \int dx \; H_u G_u^{\sim} f(x) \; g(x)$$

and therefore

$$(15.4) \qquad \text{Lim sup}_{u \uparrow \infty} \left| u \int dx \{f(x) - uG_u f(x)\} \, g(x) - u \int dx \{f(x) - uG_u^{\sim} f(x)\} \, g(x) \right|$$

$$\leq \text{Lim sup}_{u \uparrow \infty} \|f\|_{\infty} \, u \int dx \, |g(x)| \, \mathcal{E}_x \, I(X_{\zeta-0} = \partial) e^{-u\zeta} .$$

In particular if f is in $\underline{F}^{\sim}$ and if g belongs to $\underline{F}$ and is bounded with compact support

(15.5) $$E^{\sim}(f,g) = \operatorname{Lim}_{u\uparrow\infty} u \int dx \, \{f(x) - u\, G_u f(x)\}\, g(x).$$

This is true in particular for f in $\underline{\underline{F}}$ and it follows after an elementary passage to the limit that

(15.6) $$E^{\sim}(f,g) = E(f,g)$$

for f, g in $\underline{\underline{F}}$.

Remark. (15.5) does not extend to general bounded g in $\underline{\underline{F}}$. Consider for example the special case when $(\underline{\underline{F}}^{\sim}, E^{\sim}) = (\underline{\underline{F}}_a^{ref}, E)$. If $g = N\mu$ with μ bounded and if f is harmonic, then the left side of (15.5) is 0 and the right side is $\int \mu(dx) f(x)$. ///

Now we are ready to apply an argument which plays a central role in [20]. For f in $\underline{\underline{F}}^{\sim}$ and for $u > 0$

$$\int dx \, \{G_u^{\sim} f^2 - f^2 G_u^{\sim} 1\} = 0$$

and therefore

(15.7) $$\begin{aligned} u \int dx \, &\{f(x) - uG_u^{\sim} f(x)\} f(x) \\ &= u\int dx \{f(x) - uG_u^{\sim} f(x)\} f(x) + \tfrac{1}{2}u^2 \int dx \, \{G_u^{\sim} f^2(x) - f^2(x) G_u^{\sim} 1(x)\} \\ &= u \int dx \{f(x) - uG_u^{\sim} f(x)\} f(x) - \tfrac{1}{2}u \int dx \{f^2(x) - uG_u^{\sim} f^2(x)\} \\ &\qquad + \tfrac{1}{2}u \int dx \, f^2(x) \, \{1 - uG_u^{\sim} 1(x)\}. \end{aligned}$$

The point of this is that the integrand on the right can be regrouped as

(15.8) $$\begin{aligned} &\tfrac{1}{2}u^2 \{G_u^{\sim} f^2 + f^2 G_u^{\sim} 1 - 2fG_u^{\sim} f\}(x) + uf^2(x)\{1 - uG_u^{\sim} 1(x)\} \\ &= \tfrac{1}{2}u^2 \int G_u^{\sim}(x, dy)\{f(y) - f(x)\}^2 + uf^2(x)\{1 - uG_u^{\sim} 1(x)\} \end{aligned}$$

which is clearly nonnegative and indeed decreases if f is replaced by a normalized contraction f'. Therefore for f in $\underline{F}^{\sim}$, for f' a normalized contraction of f and for g satisfying $0 \leq g \leq 1$

(15.10) $$E^{\sim}(f,f) - E^{\sim}(f',f')$$

$$\geq \operatorname{Lim\,sup}_{u\uparrow\infty}[u \int dx\{f(x)-uG_u^{\sim} f(x)\} f(x)g(x)$$

$$- u\int dx \{f'(x)-uG_u^{\sim} f'(x)\} f'(x)g(x)$$

$$- \tfrac{1}{2}u \int dx\{f^2(x)-uG_u^{\sim} f^2(x)\} g(x) + \tfrac{1}{2}u \int dx\{f'^2(x)-uG_u^{\sim} f'^2(x)\} g(x)$$

$$+ \tfrac{1}{2}u\int dx \{1-uG_u^{\sim} 1(x)\} \{f^2(x)-f'^2(x)\} g(x)].$$

Suppose now that f is bounded and that g is in $\underline{F}$ and has compact support. Clearly $\psi(x)g(x)$ belongs to $\underline{F}$ for bounded ψ in domain $A^{\sim}$ and it follow after approximating f^2 by such ψ that $f^2 g$ is in $\underline{F}$ and therefore by Lemma 11.1 and (15.4)

(15.11) $$\operatorname{Lim}_{u\uparrow\infty} u \int dx\{1-uG_u^{\sim} 1(x)\} f^2(x)g(x)$$

$$= \int \varkappa(dx)f^2(x)g(x).$$

We show next that

(15.12) $$\operatorname{Lim}_{u\uparrow\infty}[u \int dx \{f(x)-u G_u^{\sim} f(x)\} f(x)g(x)$$

$$- \tfrac{1}{2}u \int dx \{f^2(x) - uG_u^{\sim}f^2(x)\} g(x)]$$

$$= \tfrac{1}{2}\int \langle A_c f\rangle (dy)\, g(y) + \tfrac{1}{2}\iint J(dy,dz)\{f(y)-f(z)\}^2 g(y)$$

$$+ \tfrac{1}{2}\int \varkappa(dy)f^2(y)g(y).$$

Note that the first term on the right is well defined since by the argument preceding (15.11) the function f is in $\underline{F}_{\ell oc}$. It follows from (15.5) and from (11.9) that (15.12) is valid for f in $\underline{F}_{(e)}$. Choose bounded f_n in $\underline{F}_{(e)}$ such that $f_n \to f$ quasi-everywhere and such that every $f_n = f$ quasi-everywhere on a fixed neighborhood D' of the support of g. By (11.9) the equation (15.12) is valid with f replaced by f_n and so (15.12) will be established if we can establish convergence of the two sides as $n \uparrow \infty$. For the right side it suffices to observe that the "error"

$$= \tfrac{1}{2}\iint_{D\times M'} J(dy, dz)[\{f(y)-f(z)\}^2 - \{f_n(y)-f_n(z)\}^2]\, g(y)$$

where $M' = \underline{X}-D'$ and where D is an open set containing the support of g and having compact closure contained in D'. It follows from (15.5) that the left side in (15.12) is unchanged if f is replaced by $f-H^M f$ in the first term and if f^2 is replaced by $f^2-H^M f^2$ in the second term. Thus convergence for the left side follows with the help of Corollary 14.3 and (15.12) is established. Thus (15.10) leads to

$$(15.13) \qquad E^{\sim}(f, f) - E^{\sim}(f', f')$$

$$\geq \tfrac{1}{2}\int\{\langle A_c f\rangle (dy) - \langle A_c f'\rangle (dy)\}\, g(y)$$

$$+ \tfrac{1}{2}\iint J(dy, dz)[\{f(y)-f(z)\}^2 - \{f'(y)-f'(z)\}^2]\, g(y)$$

$$+ \int \varkappa(dy)\{f^2(y)-f'^2(y)\}\, g(y)$$

and after passage to the limit in g

$$(15.14) \qquad E^{\sim}(f, f) - E(f, f) \geq E^{\sim}(f', f') - E(f', f').$$

This is the crucial estimate of this section. It follows from the special case $f'=0$ in (15.14) that $\underline{F}_a^{ref}$ contains $\underline{F}^{\sim}$ and that $E^{\sim}$ dominates E. Also (15.14) extends immediately to general f in $\underline{F}^{\sim}$ and (15.6) is valid for f in $\underline{F}^{\sim}_{(e)}$ and g in $\underline{F}_{(e)}$. We summarize in

Theorem 15.2. (First Structure Theorem.) Let $\tilde{P}_t$ be a strongly continuous symmetric submarkovian semigroup on $L^2(dx)$ such that the generator $\tilde{A}$ is contained in the local generator $\mathcal{O}$. Let $(\tilde{F}, \tilde{E})$ be the associated Dirichlet space.

(i) $\tilde{F}$ contains the given Dirichlet space F and is contained in the active reflective space F_a^{ref}.

(ii) If f is in the extended space $F_{(e)}$, then Hf is also in $\tilde{F}_{(e)}$, the difference f-Hf is in $F_{(e)}$ and

(15.15) $$\tilde{E}(f, f) = E(f-Hf, f-Hf) + \tilde{E}(Hf, Hf).$$

Equivalently, (15.6) is valid for f in $\tilde{F}_{(e)}$ and g in $F_{(e)}$. Also for $u > 0$ the operator H_u implements $\tilde{E}_u$ orthogonal projection of $\tilde{F}$ onto the complement of F.

(iii) The difference $\tilde{E}$-E is contractive on $\tilde{F}_{(e)}$. That is, (15.14) is valid for f in $\tilde{F}_{(e)}$ and for f′ a normalized contraction of f. ///

16. The Recurrent Case

In this section we treat the case when $(\underline{F}, E)$ is recurrent. We will see in particular that $\underline{F}^{ref} = \underline{F}_{(e)}$ and that the First Structure Theorem collapses.

We begin by establishing the recurrent analogue of Theorem 13.1. Let D, M be as in Section 13 and again we restrict attention to functions having a representation (13.1) with h satisfying (13.4). Only now the variable Φ is not present. Fix $u > 0$ and let R be the usual terminal variable exponentially distributed at the rate u and independent of the trajectory variables. As noted in Section 8 the process $u \int dx\, \theta_x$ plays the role of the approximate Markov process of Section 5 if R is interpreted as a "death time." Clearly the functionals $Mf(t)$, $M_c f(t)$, $\langle Mf\rangle(dt)$, $\langle M_c f\rangle(dt)$ are well defined at least for t such that $X_t \in D$ or $X_{t-0} \in D$. Here we represent the random set

$$\{t : \sigma(M) < t < R,\ X_t \in D \text{ and } X_{t-0} \in D\}$$

as a union of intervals $(e_u(i), r_u(i))$. The "preliminary formula" to be established is

Theorem 16.1. Assume that $(\underline{F}, E)$ is recurrent and let f have a representation (13.1) with h satifying (13.2) and (13.4), except that Φ is never present. Then for $u > 0$

$$(16.1)\quad \tfrac{1}{2}u \int dx\, \mathscr{S}_x \int_0^R \langle Mf\rangle(dt)\, 1_D(X_t) + \tfrac{1}{2}u \int dx\, \mathscr{S}_x\, \Sigma_i \{f(X_{e_u(i)}) - f(X_{e_u(i)-0})\}^2$$

$$= E(f_0, f_0) + \tfrac{1}{2}u \int_D dx\, \mathscr{S}_x \{h(X_{\sigma(M)\wedge R}) - h(X_0)\}^2$$

$$+ \tfrac{1}{2}u \int dx\, \mathscr{S}_x\, \Sigma_i \{h(X_{r_u(i)}) - h(X_{e_u(i)-0})\}^2$$

Equivalently

$$(16.1') \quad \tfrac{1}{2}u \int dx\, \mathcal{E}_x \int_0^R \langle M_c f\rangle (dt) 1_D(X_t) + \iint_{D\times M} J(dy, dz)\, \{f(y)-f(z)\}^2$$

$$+ \tfrac{1}{2}\iint_{D\times D} J(dy, dz)\{f(y)-f(z)\}^2$$

$$= E(f_o, f_o) + \tfrac{1}{2} u \int_D dx\, \mathcal{E}_x \{h(X_{\sigma(M)\wedge R}) - h(X_0)\}^2$$

$$+ \tfrac{1}{2}u \int dx\, \mathcal{E}_x \Sigma_i \{h(X_{r_u(i)}) - h(X_{e_u(i)-0})\}^2 \quad ///$$

Proof. It suffices to consider the special case when h is bounded and when $f_o = G^D\varphi$ with both φ and f_o bounded and integrable. (Note that such f_o are dense in $\underline{\underline{F}}^D_{(e)}$ since if ψ is bounded and integrable then $f_o = G_1^D \psi = G^D(\psi - f)$ qualifies.) First

$$(16.2) \qquad \tfrac{1}{2}\mathcal{E}_x \int_0^{\sigma(M)\wedge R} \langle Mf\rangle (dt)$$

$$= \tfrac{1}{2}\, \mathcal{E}_x \{f(X_{\sigma(M)\wedge R}) - f(X_0) + \int_0^{\sigma(M)} dt\, \varphi(X_t)\}^2$$

$$= \mathcal{E}_x \int_0^{\sigma(M)\wedge R} dt\, \varphi(X_t)\{\int_t^{\sigma(M)\wedge R} ds\,\varphi(X_s) + f_o(X_{\sigma(M)\wedge R})\}$$

$$+ \tfrac{1}{2}\, \mathcal{E}_x\{f(X_{\sigma(M)\wedge R}) - f(X_0)\}^2$$

$$+ \mathcal{E}_x\{h(X_{\sigma(M)\wedge R}) - f(X_0)\} \int_0^{\sigma(M)\wedge R} dt\, \varphi(X_t)$$

$$= \mathcal{E}_x \int_0^{\sigma(M)\wedge R} dt\, \varphi(X_t) f_o(X_t)$$

$$+ \tfrac{1}{2}\mathcal{E}_x\{h(X_{\sigma(M)\wedge R}) - h(X_0)\}^2 + \tfrac{1}{2}\, \mathcal{E}_x\{f_o(X_{\sigma(M)\wedge R}) - f_o(X_0)\}^2$$

$$+ \mathcal{E}_x\{h(X_{\sigma(M)\wedge R}) - h(X_0)\}\{f_o(X_{\sigma(M)\wedge R}) - f_o(X_0)\}$$

$$+ \mathcal{E}_x\{h(X_{\sigma(M)\wedge R}) - h(X_0)\} \int_0^{\sigma(M)\wedge R} dt\, \varphi(X_t)$$

$$- \mathcal{E}_x\, f_o(X_0) \int_0^{\sigma(M)\wedge R} dt\, \varphi(X_t).$$

Next fix D' with compact closure contained in D and let

(16.3) $\quad \epsilon_u(1) = \inf\{t > \sigma(M) \colon X_t \in D' \text{ and } t < R\}$

$\qquad \rho_u(1) = \inf\{t > \epsilon(1) \colon X_t \in M\} \wedge R$

etc.

with the usual understanding that these times are $+\infty$ when not otherwise defined. For each i

$$(16.4)\qquad \tfrac{1}{2}\,\mathcal{E}_x \int_{\epsilon_u(i)}^{\rho_u(i)} \langle Mf\rangle\,(dt)$$

$$= \tfrac{1}{2}\,\mathcal{E}_x \{f(X_{\rho_u(i)}) - f(X_{\epsilon_u(i)}) + \int_{\epsilon_u(i)}^{\rho_u(i)} dt\,\varphi(X_t)\}^2$$

$$= \mathcal{E}_x \int_{\epsilon_u(i)}^{\rho_u(i)} dt\,\varphi(X_t)\{\int_t^{\rho_u(i)} ds\,\varphi(X_s) + f_0(X_{\rho_u(i)})\}$$

$$+ \tfrac{1}{2}\,\mathcal{E}_x \{f(X_{\rho_u(i)}) - f(X_{\epsilon_u(i)})\}^2$$

$$+ \mathcal{E}_x\{h(X_{\rho_u(i)}) - f(X_{\epsilon_u(i)})\}\{\int_{\epsilon_u(i)}^{\rho_u(i)} dt\,\varphi(X_t).$$

Assume temporarilly that the intervals $(e_u(i), r_u(i))$ are labeled so that

(16.3') $\quad e_u(i) = \sup\{t < \epsilon_u(i) \colon X_t \text{ or } X_{t-0} \in M\}$

Whenever $\epsilon_u(i) < +\infty$. For each i

$$\mathcal{E}_x\{f(X_{\epsilon_u(i)}) - f(X_{e_u(i)-0})\}\{f(X_{\rho_u(i)}) - f(X_{\epsilon_u(i)}) + \int_{\epsilon_u(i)}^{r_u(i)} dt\,\varphi(X_t)\} = 0$$

and so (16.4) can be rewritten

$$(16.5)\qquad \tfrac{1}{2}\mathscr{E}_{\mathbf{x}} \int_{\epsilon_u(i)}^{\rho_u(i)} \langle Mf\rangle(dt)$$

$$= \mathscr{E}_{\mathbf{x}} \int_{\epsilon_u(i)}^{\rho_u(i)} dt\ \varphi(X_t)\, f_0(X_t)$$

$$+ \tfrac{1}{2}\, \mathscr{E}_{\mathbf{x}}\{f(X_{\rho_u(i)}) - f(X_{e_u(i)-0})\}^2$$

$$- \tfrac{1}{2}\, \mathscr{E}_{\mathbf{x}}\{f(X_{\epsilon_u(i)}) - f(X_{e_u(i)-0})\}^2$$

$$+ \mathscr{E}_{\mathbf{x}}\{h(X_{\rho_u(i)}) - h(X_{e_u(i)-0})\} \int_{\epsilon_u(i)}^{\rho_u(i)} dt\ \varphi(X_t).$$

Summing over i and passing to the limit as in Section 13, we get

$$(16.6)\qquad \tfrac{1}{2}\mathscr{E}_{\mathbf{x}}\, I(\sigma(M) < R) \int_{\sigma(M)}^{R} \langle Mf\rangle\,(dt)\, 1_D(X_t)$$

$$+ \tfrac{1}{2}\, \mathscr{E}_{\mathbf{x}}\, \Sigma_i\{f(X_{e_u(i)}) - f(X_{e_u(i)-0})\}^2$$

$$= \mathscr{E}_{\mathbf{x}} I(\sigma(M) < R) \int_{\sigma(M)}^{R} dt\ \varphi(X_t)\, f_0(X_t)$$

$$+ \tfrac{1}{2}\, \mathscr{E}_{\mathbf{x}} \Sigma_i\{f(X_{\rho_u(i)}) - f(X_{e_u(i)-0})\}^2$$

$$+ \mathscr{E}_{\mathbf{x}} \Sigma_i\{h(X_{\rho_u(i)}) - h(X_{e_u(i)-0})\} \int_{e_u(i)}^{\rho_u(i)} dt\ \varphi(X_t).$$

After combining with (16.5), integrating with respect to udx and eliminating a cross term with the help of time reversal as in Section 13 we get

$$(16.7)\quad \tfrac{1}{2}u\int dx\,\mathcal{E}_{\mathbf{x}}\int_0^R \langle Mf\rangle(dt)\,1_D(X_t)+\tfrac{1}{2}u\int dx\,\mathcal{E}_{\mathbf{x}}\Sigma_i\{f(X_{e_u(i)})-f(X_{e_u(i)-0})\}^2$$

$$= E(f_o,f_o)+\tfrac{1}{2}u\int dx\,\mathcal{E}_{\mathbf{x}}\{h(X_{\sigma(M)\wedge R})-h(X_0)\}^2$$

$$+\tfrac{1}{2}u\int dx\,\mathcal{E}_{\mathbf{x}}\,\Sigma_i\{h(X_{r_u(i)})-h(X_{e_u(i)-0})\}^2$$

$$+\tfrac{1}{2}u\int dx\,\mathcal{E}_{\mathbf{x}}\,\{f_o(X_{\sigma(M)\wedge R})-f_0(X_0)\}^2$$

$$+\tfrac{1}{2}u\int dx\,\mathcal{E}_{\mathbf{x}}\Sigma_i I(r_u(i)=R)f^2(X_R)$$

$$-u\int_D dx\,f_o(x)\,G_u^D\,\varphi(x)$$

$$+u\int dx\,\mathcal{E}_{\mathbf{x}}\{f_o(X_{\sigma(M)\wedge R})-f_o(X_0)\}\{h(X_{\sigma(M)\wedge R})-h(X_0)\}$$

$$+u\int dx\,\mathcal{E}_{\mathbf{x}}\Sigma_i I(r_u(i)=R\}\;f_o(X_R)\{h(X_R)-h(X_{e_u(i)-0})\}\,.$$

The theorem follows since

$$\tfrac{1}{2}u\int dx\,\mathcal{E}_{\mathbf{x}}\{f_o(X_{\sigma(M)\wedge R})-f_o(X_0)\}^2-u\int_D dx\,f_o(x)\,G_u^D\,\varphi(x)$$

$$+\tfrac{1}{2}u\int dx\,\mathcal{E}_{\mathbf{x}}\Sigma_i\,I(r_u(i)=R)\,f_o^2(X_R)$$

$$=\tfrac{1}{2}u\int_D dx\,f_o^2(x)\,H_u^M 1(x)+u^2\int_D dx\,\{G_u^D f_o^2(x)-f_o(x)G_u^D f_o(x)\}$$

$$-u\int_D dx\,f_o(x)\,\{f_o(x)-u\,G_u^D\,f_o(x)\}$$

$$+\tfrac{1}{2}u\int_D dx\,f_o^2(x)\,H_u^M 1(x)$$

$$=0$$

and since

$$u\int dx\,\mathcal{E}_{\mathbf{x}}\{f_o(X_{\sigma(M)\wedge R})-f_o(X_0)\}\{h(X_{\sigma(M)\wedge R})-h(X_0)\}$$

$$+u\int dx\,\mathcal{E}_{\mathbf{x}}\,\Sigma_i\,I(r_u(i)=R)f_o(X_R)\{h(X_R)-h(X_{e_u(i)-0})\}$$

$$= u \int_D dx\, \mathcal{E}_x I(R < \sigma(M))\, f_0(X_R)\, \{h(X_R) - h(X_0)\}$$

$$+ u \int_D dx\, \mathcal{E}_x f_0(X_0) I(\sigma(M) < R)\{h(X_{\sigma(M)}) - h(X_0)\}$$

$$= u \int_D dx\, \mathcal{E}_x f_0(X_0)\{h(X_{R \wedge \sigma(M)}) - h(X_0)\}$$

$$= 0. \; ///$$

The local Dirichlet space $\underline{F}_{\ell oc}$ and the reflected space $\underline{F}^{ref}$ are defined as for the transient case. Also Lemma 14.1 is applicable. If $\underline{X}$ is compact then it follows directly from Lemma 14.1 that $\underline{F}_{(e)}$ contains every bounded function in $\underline{F}^{ref}$ and therefore all of $\underline{F}^{ref}$ When $\underline{X}$ is not compact the proof of Theorem 14.2 shows that if f belongs to $\underline{F}^{ref}$ and if D is open with compact closure, then $f - H^M f$ belongs to $\underline{F}_{(e)}$ and $E(f - H^M f, f - H^M f) \le E(f, f)$. If f is bounded then by Theorem 8.10-(ii) there exist $D_n \uparrow \underline{X}$ such that the Cesàro sums $(1/n)\{H^{M_1} f + \cdots + H^{M_n} f\}$ converge quasi-everywhere and it follows that $f = f_0 + h$ with f_0 in $\underline{F}_{(e)}$ and with h bounded and harmonic. But since 1 is in $\underline{F}_{(e)}$ it follows from the $u > 0$ version of Lemma 3.3 (see the proof of Theorem 8.7) that $\underline{F}_{(e)}$ contains h and therefore f. The restriction to f bounded is easily removed and we have proved

<u>Theorem 16.2.</u> <u>If</u> $(\underline{F}, E)$ <u>is recurrent, then</u> $\underline{F}^{ref} = \underline{F}_{(e)}$. ///

Let $\tilde{P}_t$, $t > 0$ be a submarkovian symmetric semigroup on $L^2(dx)$ with generator $\tilde{A}$ contained in the local generator $\mathcal{A}$ and let $\tilde{G}_u$, $u > 0$ be the corresponding resolvent. The paragraph preceeding (15.3) shows that for φ bounded and for $u > 0$ the difference $f = \tilde{G}_u \varphi - G_u \varphi$ is u-harmonic. But also f is bounded and since (F, E) is recurrent it follows that necessarily $f = 0$. Thus $\tilde{G}_u = G_u$ and so

Theorem 16.3. *If $(\underline{F}, E)$ is recurrent, then $P_t, t > 0$ is the only symmetric submarkovian semigroup on $L^2(dx)$ with generator containd in the local generator $\mathfrak{a}$.* ///

We finish with

Theorem 16.4. *If $(\underline{F}, E)$ is recurrent and irreducible, then every harmonic function h in the extended space $\underline{F}_{(e)}$ is constant.* ///

Proof. For h bounded the theorem follows directly from Proposition 4.16 and the convergence theorem for bounded martingales. For general h we need a more subtle argument. For $u > 0$

$$E(h,h) = u \int dx\, \mathcal{E}_x \int_0^R \langle Mf \rangle (dt)$$

$$= u \int dx\, \mathcal{E}_x \{h(X_R) - h(X_0)\}^2$$

with R as above. It follows in particular that for almost every x

$$\sup_{t \geq 0} \mathcal{E}_x\, h^2(X_{t \wedge R}) < +\infty$$

and so by the convergence theorem for L^2 bounded martingales $h = uG_u h$. Clearly harmonicity is unaffected by random time change and so by Theorem 8.5 we can assume that h is in $L^2(dx)$. But then Lemma 1.1 is applicable and $E(h,h) = 0$. Finally $(1/t) \int dx \int P_t(x,dy)\, \{h(x)-h(y)\}^2 = 0$ for $t > 0$ by Lemma 1.7-(ii) which contradicts irreducibility unless h is constant. ///

17. Scope of First Structure Theorem

In this section we look briefly at the special case of stable Markov chains, primarily in order to indicate the limitations of the First Structure Theorem. Our main result is that the First Structure Theorem is applicable if and only if the appropriate Kolmogorov equation is satisfied

Let $\underline{I}$ be a denumerably infinite set and let $\tilde{P}_t(x,y)$ be a standard transition matrix on I. That is, $\tilde{P}_t(x,y)$ is defined for $t > 0$ and for x,y in $\underline{I}$ and satisfies

17.1.1. $\tilde{P}_t(x,y) \geq 0$; $\Sigma_y \tilde{P}_t(x,y) \leq 1$.

17.1.2. $\Sigma_z \tilde{P}_t(x,z) \tilde{P}_s(z,y) = \tilde{P}_{t+s}(x,y)$.

17.1.3. $\tilde{P}_t(x,y) \to \delta_x(y)$ as $t \downarrow 0$. ///

We are using ϵ_x to denote the point mass at x. Thus 17.1.3 is equivalent to

$$\tilde{P}_t(x,x) \to 1 \; ; \; \tilde{P}_t(x,y) \to 0 \qquad \text{for } x \neq y.$$

To maintain a formal distinction between measures and functions we introduce a separate symbol e_x for the indicator of the set $\{x\}$.

We restrict attention to transition matrices satisfying:

17.2. Condition of Symmetry. There exists an everywhere positive measure $m(x)$ on $\underline{I}$ such that

$$m(x)\, \tilde{P}_t(x,y) = m(y)\, \tilde{P}_t(y,x). \; /// \tag{17.1}$$

The operators $\tilde{P}_t$ defined by

$$\tilde{P}_t f(x) = \Sigma_y P_t(x,y) f(y)$$

form a strongly continuous symmetric submarkovian semigroup on $L^2(\underline{\underline{I}},m)$. Let $(\tilde{\underline{\underline{F}}},\tilde{E})$ be the corresponding Dirichlet space. Since

$$m(x)\{1-\tilde{P}_t(x,x)\} = \Sigma_y m(y)\,(1-\tilde{P}_t e_x(y))\,e_x(y)$$

it follows from Lemma 1.1 that

(17.2) $$q(x) = \text{Lim}_{t\downarrow 0}\,(1/t)\{1-\tilde{P}_t(x,x)\}$$

exists for x in $\underline{\underline{I}}$ with $0 \leq q(x) \leq +\infty$. The state x is said to be <u>stable</u> if $q(x) < +\infty$ and <u>instaneous</u> if $q(x) + \infty$. We assume here that every state is stable and define

(17.3) $$\alpha(x) = m(x)q(x).$$

Then e_x belongs to $\tilde{\underline{\underline{F}}}$ and

(17.4) $$\tilde{E}(e_x,e_x) = \alpha(x).$$

Also Lemma 1.1 guarantees that

(17.5) $$q(x)P(x,y) = \text{Lim}_{t\downarrow 0}\,(1/t)\,\tilde{P}_t(x,y) \qquad x \neq y$$

exists and is finite for x,y in $\underline{\underline{I}}$. (Of course the existence of these limits can be deduced by other techniques without using 17.2. See [5].) We establish the convention

(17.5') $$P(x,x) = 0$$

and then it is easy to check that P is substochastic:

(17.6) $$P(x,y) \geq 0; \quad \Sigma_y P(x,y) \leq 1$$

and that α symmetrizes P:

(17.7) $$\alpha(x)\,P(x,y) = \alpha(y)\,P(y,x).$$

Next we impose

17.3. Condition of Irreducbility. The rate $q(x) > 0$ for all x and the road map P is irreducible. ///

There is no real loss of generality in this.

17.4. Definition. f in $L^2(m)$ belongs to the domain of the local generator Ω if Pf converges absolutely on $\underline{I}$ and if $q(Pf-f)$ is in $L^2(m)$. In this case

(17.8) $$\Omega f(x) = q(x)Pf(x) - q(x)f(x). \quad ///$$

Theorem 17.1. The following are equivalent for the standard transition matrix $P_t^\sim(x,y)$.

(i) The generator $A^\sim$ of the semigroup $P_t^\sim, t > 0$ is contained in the local generator Ω.

(ii) $P_t^\sim(x,y)$ satisfies the backward Kolmogorov equation

(17.9) $$(d/dt)\, P_t^\sim(x,y) = -\, q(x)\, P_t^\sim(x,y) + q(x)\Sigma_z\, P(x,z)\, P_t^\sim(z,y)$$

or, equivalently

(17.9') $$P_t^\sim(x,y) = e^{-tq(x)}\, e_x(y) + \int_0^t ds\; q(x)e^{-sq(x)}\, \Sigma_z P(x,z)\, P_{t-s}^\sim(z,y).$$

(iii) $P_t^\sim(x,y)$ satisfies the forward Kolmogorov equation

(17.10) $$(d/dt)\, P_t^\sim(x,y) = -P_t^\sim(x,y)q(y) + \Sigma_z P_t^\sim(x,z)q(z)P\,(z,y)$$

or, equivalently

(17.10') $$P_t^\sim(x,y) = e^{-tq(y)}\, e_y(x) + \int_0^t ds\; \Sigma_z\, P_s^\sim(x,z)q(z)P(z,y)e^{-(t-s)q(y)} \quad . \;///$$

Proof. If the generator $A^{\sim}$ is contained in the local generator Ω, then

$$(17.11) \qquad \tilde{E}(e_x, f) = -\Omega f(x)$$

for f in domain $A^{\sim}$. Any bounded f in $\underline{F}^{\sim}$ can be approximated both pointwise and in the $\tilde{E}$ sense by uniformly bounded functions in domain $A^{\sim}$ and so (17.11) is valid also for f in $\underline{F}^{\sim}_b$ and therefore general f in $\underline{F}^{\sim}$. In particular (17.11) is valid for $f = P^{\sim}_t e_y$ and (17.9) follows with the help of Lemma 1.1. Thus (i) implies (ii). Conversely if (ii) is true then (17.11) is valid whenever $f = P_t e_y$. But it is easy to check that the linear span of such functions is dense in $\tilde{\underline{F}}$. Thus (17.11) is valid for general f in $\underline{F}^{\sim}$ and in particular for f in domain $A^{\sim}$ and (i) follows. Finally equivalence of (ii) and (iii) follows directly from symmetry. ///

Let $P_t^{min}(x,y)$ be the well known minimal transition matrix of W.Feller [15]. This is defined by

$$P_t^{(0)}(x,y) = e^{-tq(x)} e_x(y)$$

$$P_t^{(n+1)}(x,y) = \int_0^t ds\, q(x) e^{-sq(x)}\, P\, P_{t-s}(x,y)$$

$$P_t^{min}(x,y) = \Sigma_{n=0}^{\infty} P_t^{(n)}(x,y).$$

It is easy to check that $P_t^{min}(x,y)$ satisfies 17.1 and 17.2 and also the three equivalent conditions of Theorem 17.1. Also it is important that $P_t^{min}(x,y)$ can be characterized as the minimal nonnegative solution of (17.9') or (17.10'). Let $\mathcal{E}$ be defined by

$$(17.12) \qquad \mathcal{E}(f,f) = \Sigma_x \alpha(x) \{1 - P1(x)\} f^2(x) + \frac{1}{2} \Sigma_{x,y} \alpha(x) P(x,y) \{f(x) - f(y)\}^2$$

when it converges, let $\mathcal{J}$ be the collection of functions f on $\underline{\underline{I}}$ for which (17.12) converges and let $\mathcal{T}^{min}$ be the set of f in $\mathcal{T}$ for which there exists $f_n, n \geq 1$ with finite support such that

$$f_n \to f \text{ on } \underline{\underline{I}}\, ; \quad \sup_n \mathcal{E}(f_n, f_n) < +\infty .$$

Also let $\underline{\underline{F}}^{min} = \mathcal{T}^{min} \cap L^2(m)$.

<u>Theorem 17.2.</u> <u>$(\underline{\underline{F}}^{min}, \mathcal{E})$ is the Dirichlet space on $L^2(m)$ which corresponds to the minimal semigroup P_t^{min}, $t > 0$. Also $(\underline{\underline{F}}^{min}, \mathcal{E})$ is regular on $\underline{\underline{I}}$ and</u>

$$\underline{\underline{F}}^{min}_{(e)} = \mathcal{T}^{min} \; ; \; \underline{\underline{F}}^{min, ref} = \mathcal{T} . \; ///$$

<u>Proof.</u> Let $(\underline{\underline{F}}^o, E^o)$ be the Dirichlet space associated with $P_t^{min}, t > 0$. Since P_t^{min} satisfies (17.2) and (17.5), the space $\underline{\underline{F}}^{min}$ is contained in $\underline{\underline{F}}^o$ and $E^o(f,g) = \mathcal{E}(f,g)$ for f,g in $\underline{\underline{F}}^{min}$. Thus for the first sentence it suffices to show that actually $\underline{\underline{F}}^o$ is contained in $\underline{\underline{F}}^{min}$. But the semigroup $P_t^o, t > 0$ associated with $(\underline{\underline{F}}^{min}, \mathcal{E})$ satisfies (17.9'), (17.10') and since P_t^{min} is the minimal solution, $P_t^o \geq P_t^{min}$. But then $\underline{\underline{F}}^o$ is contained in $\underline{\underline{F}}^{min}$ and the first sentence is proved. The remainder of the theorem is clear since

$$J(x,y) = \alpha(x)\, P(x,y) \; ; \; \kappa(x) = \alpha(x)\, \{1 - P1(x)\} \; ; \; D = 0 . \; ///$$

To make connection with the theory in Section 15 we need

<u>Lemma 17.3.</u> (i) <u>A function h is harmonic for $(\underline{\underline{F}}^{min}, \mathcal{E})$ in the sense of Definition 14.3 if and only if Ph converges absolutely and</u> h = Ph.

(ii) <u>The local generator $\mathfrak{a}$ for $(\underline{\underline{F}}^{min}, \mathcal{E})$ in the sense of Definition 15.1 is identical with the local generator</u> Ω . ///

<u>Proof</u> (i) follows easily from the well known interpretation of P as the "road map" for the minimal process and then (ii) follows directly from (i). ///

It follows directly that the First Structure Theorem is applicable to the standard transition matrix $P_t^{\sim}(x,y)$ if and only if it satisfies the Kolmorogov equations (17.9), (17.10).

We finish by constructing an example with state space $[0,1]$ which does not fit into the framework of the First Structure Theorem and then we apply random time change to obtain an example of a stable Markov chain which does not satisfy the Kolmogorov equations.

Let $\underline{X}$ be the open interval $(0,1)$ and let $\kappa(dt)$ be any nontrivial bounded Radon measure on $(0,1)$. (This requirement that $\kappa(dt)$ be bounded is imposed only to avoid ambiguity about boundary conditions. It has no other significance in the present context.) Let $\underline{\underline{F}}$ be the collection of absolutely continuous functions f on the closed interval $[0,1]$ which satisfy

$$\int_0^1 dt\,\{f'(t)\}^2 < +\infty \tag{17.13}$$

and in addition the boundary conditions

$$f(0) = 0\ ;\quad f(1) = 0. \tag{17.14}$$

For f in $\underline{\underline{F}}$ let

$$E(f,f) = \tfrac{1}{2}\int_0^1 dt\,\{f'(t)\}^2 + \int \kappa(dt)\,f^2(t). \tag{17.15}$$

Then $(\underline{\underline{F}},E)$ is the Dirichlet space on $L^2(\underline{\underline{X}},dt)$ which corresponds to Brownian motion with an absorbing barrier and with "killing at the rate κ." (See [28] for precise definitions.) The reflected space $\underline{\underline{F}}^{ref}$ is the collection of absolutely continuous f on $[0,1]$ satisfying (17.13) only. For f in $\underline{\underline{F}}_{ref}$ let

$$E^{\sim}(f,f) = \tfrac{1}{2}\int_0^1 dt\,\{f'(t)\}^2 + \int \kappa(dt)\,\{f(t)-f(0)\}^2. \tag{17.16}$$

Then $(\underline{\underline{F}}^{ref}, E^{\sim})$ is a Dirichlet space on $L^2(dx)$ which is regular on the closed interval $[0,1]$. It corresponds to reflecting barrier Brownian motion on which is "superimposed" an intensity for jumping to and away from 0 at the "rate" $\varkappa(dt)$. (This heuristic description can be justified in the context of Chapter II.) Clearly $(\underline{\underline{F}}, E)$ corresponds to the process obtained from $(\underline{\underline{F}}^{ref}, E^{\sim})$ by absorbtion upon exiting from $(0,1)$. Let $\mathcal{A}$ be the local generator for $(\underline{\underline{F}}, E)$. With the techniques of [28] it is easy to check that f belongs to the domain of $\mathcal{A}$ and $\mathcal{A}f = \varphi$ if and only if the derivative f' is in bounded variation and $\frac{1}{2}f'(dt) - f(t)\varkappa(dt) = \varphi(t)\,dt$ as signed measures on $(0,1)$. If the generator $A^{\sim}$ were contained in $\mathcal{A}$, then it would follow from (15.15) that

$$E^{\sim}(f,g) = \frac{1}{2}\int_0^1 dt\, f'(t)g'(t) + \int \varkappa(dt) f(t) g(t) \tag{17.17}$$

for f in $\underline{\underline{F}}^{ref}$ and g in $\underline{\underline{F}}$ which clearly contradicts (17.16). Thus the First Structure Theorem is not applicable to $(\underline{\underline{F}}^{ref}, E^{\sim})$.

Let $\underline{\underline{I}}$ be a denumerably infinite subset if $(0,1]$ which contains 1 and whose only limit is 0 and let $\nu(x)$ be a bounded measure on $\underline{\underline{I}}$ which charges every point of $\underline{\underline{I}}$. Let $(\underline{\underline{F}}^{\nu}, E^{\nu})$ be the time changed Dirichlet space of Section 8 with $(\underline{\underline{F}}^{ref}, E^{\sim})$ playing the role of $(\underline{\underline{F}}, E)$ in Section 8. The associated semigroup P_t^{ν} comes from a standard transition matrix $P_t^{\nu}(x,y)$ on $\underline{\underline{I}}$. For f defined on $\underline{\underline{I}}$ let Lf be the unique extension to $(0,1]$ which satisfies the homogeneous equation

$$Lf'(dx) = Lf(x)\,\varkappa(dx) \tag{17.18}$$

on the complement of $\underline{\underline{I}}$. It follows from the results in Section 8 that f belongs to the time changed space $\underline{\underline{F}}^{\nu}$ if and only if Lf belongs to $\underline{\underline{F}}_{ref}$ and then

$$E^{\nu}(f,f) = E^{\sim}(Lf, Lf).$$

17.8

In particular the indicators e_x, x in $\underline{\underline{I}}$ belong to $\underline{\underline{E}}^\nu$ and so every state is stable. Clearly P and α depend on $\underline{\underline{I}}$. However we can be sure that $P(x,y) > 0$ only for adjacent elements in $\underline{\underline{I}}$. After comparing with the corresponding time changed absorbed process, it is easy to see that for x,y in $\underline{\underline{I}}$ and for $t > 0$

$$E^\nu(e_x, P_t^\nu e_y) = \Sigma_z \alpha(x)\, P(x,z)\, \{P_t^\nu(x,y) - P_t^\nu(z,y)\}$$

$$+ \alpha(x)\, \{1 - P1(x)\}\, \{P_t^\nu(x,y) - LP_t^\nu e_y(0)\}\,.$$

Thus

$$(d/dt)\, \tilde{P}_t(x,y) = -q(x)\, \tilde{P}_t(x,y) + q(x)\, \Sigma_z P(x,z)\, \tilde{P}_t(z,y)$$

$$+ q(x)\, \{1 - P1(x)\}\, LP_t^\nu e_y(0)$$

and $\tilde{P}_t(x,y)$ does not satisfy the Kolmogorov equations since the last term cannot vanish for all choices of $t > 0$ and of x,y in $\underline{\underline{I}}$.

18. Enveloping Dirichlet Space

In this section we assume that $(\underline{F},E)$ is transient and that the killing measure $\kappa(dx)$ is nontrivial.

To avoid excessive verbiage below we introduce here a special notation for truncations:

(18.1) $$\tau_n f(x) = \begin{cases} f(x) & \text{for } |f(x)| \leq n \\ n \operatorname{sgn} f(x) & \text{for } |f(x)| > n. \end{cases}$$

Obviously each τ_n is a normalized contraction on R (that is, satisfies (1.12)).

For f in $\underline{F}_{loc}$ we define the resurrected form

(18.2) $$E^{res}(f,f) = \tfrac{1}{2}\iint J(dy,dz)\{f(y)-f(z)\}^2 + \tfrac{1}{2}\int \langle A_c f\rangle (dy)$$

18.1. Definition. A function f belongs to the enveloping Dirichlet space $\underline{F}^{env}$ if it satisfies the following conditions:

18.1.1. f is defined and finite quasi-everywhere.

18.1.2. The truncations $\tau_n(f)$, $n \geq 1$ all belong to the local space $\underline{F}_{loc}$.

18.1.3. $\sup_n E^{res}(\tau_n(f),\tau_n(f)) < +\infty$. ///

Clearly E^{res} is contractive and it follows that

$$E^{res}(f,f) = \operatorname{Lim}_{n\uparrow\infty} E^{res}(\tau_n(f),\tau_n(f))$$

is well defined for f in $\underline{F}_{env}$. Also

$$\langle A_c f\rangle (dy) = \operatorname{Lim}_{n\uparrow\infty} \langle A_c \tau_n(f)\rangle (dy)$$

is defined and (18.2) is valid.

18.2. Definition. A function f in the enveloping Dirichlet space $\underline{\underline{F}}^{env}$ belongs to the resurrected Dirichlet space $\underline{\underline{F}}^{res}$ if there exists a sequence $f_n, n \geq 1$ in the extended space $\underline{\underline{F}}_{(e)}$ such that

18.2.1. $\{f_n\}$ is Cauchy relative to E^{res}.

18.2.2. $f_n \to f$ quasi - everywhere. ///

As for 1.6.1 the condition 18.2.1 can be replaced by

18.2.1'. $E^{res}(f_n, f_n)$ is bounded independent of n.

As in Section 14 for the reflected space we introduce also the active spaces

$$(18.3) \qquad \underline{\underline{F}}_a^{env} = \underline{\underline{F}}^{env} \cap L^2(dx); \ \underline{\underline{F}}_a^{res} = \underline{\underline{F}}^{res} \cap L^2(dx).$$

Theorem 18.1. (i) The active resurrected space $(\underline{\underline{F}}_a^{res}, E^{res})$ is a regular Dirichlet space on $L^2(dx)$.

(ii) A set A is polar for $\underline{\underline{F}}_a^{res}$ if and only if it is polar for $\underline{\underline{F}}$ and a function f is quasi-continuous for $\underline{\underline{F}}_a^{res}$ if and only if it is quasi-continuous for $\underline{\underline{F}}$.

(iii) $\underline{\underline{F}}^{res}$ is the extended Dirichlet space for $\underline{\underline{F}}_a^{res}$.

(iv) $\underline{\underline{F}}^{env}$ is the reflected space for $(\underline{\underline{F}}_a^{res}, E^{res})$. ///

Proof. If the reference measure dx dominates killing measure $\kappa(dx)$ then $\underline{\underline{F}}_a^{res} = \underline{\underline{F}}$ and the theorem for this case follows easily with the help of Propositions 3.20 and 3.22. The general case then follows from the results in Section 8. ///

The adjective "resurrected" is justified by

<u>Theorem 18.2.</u> (i) <u>For $\varphi \geq 0$ the functions $P_t^{res}\varphi$ and $G_u^{res}\varphi$ are the minimal nonnegative solutions of</u>

(18.4) $$P_t^{res}\varphi(x) = P_t\varphi(x) + \mathscr{E}_x I(\zeta \leq t;\ X_{\zeta-0} \varepsilon \underline{X})\, P_{t-\zeta}^{res}\varphi(X_{\zeta-0})$$

(18.4') $$G_u^{res}\varphi(x) = G_u\varphi(x) + \mathscr{E}_x e^{-u\zeta} I(X_{\zeta-0} \varepsilon \underline{X})\, G_u^{res}\varphi(X_{\zeta-0})$$

(ii) <u>The transition operators P_t can be recovered from the resurrected process by</u>

(18.5) $$P_t\varphi(x) = \mathscr{E}_x^{res} e^{-a(\varkappa;t)} \varphi(X_t) \quad ///$$

<u>Proof.</u> The results on random time change in Section 8 permit us to restrict attention to the special case when $\varkappa(dx)$ is absolutely continuous with bounded density $\varkappa(x)$ so that $\underline{F}^{res} = \underline{F}$. For (i) it suffices to establish (18.4'). A routine computational argument establishes the existence of a symmetric submarkovian resolvent $\tilde{G}_u, u > 0$ such that $\tilde{G}_u\varphi$ is the minimal nonnegative solution of

(18.6) $$\tilde{G}_u\varphi = G_u\varphi + G_u\varkappa\tilde{G}_u\varphi$$

and therefore also of

(18.6') $$\tilde{G}_u\varphi = G_u\varphi + \tilde{G}_u\varkappa G_u\varphi .$$

But it follows directly from these equations that $(\underline{F}_a^{res}, E^{res})$ is the associated Dirichlet space and (i) is proved. For (ii) let P_t^o be defined by (18.5). That the P_t^o form a submarkovian semigroup follows from a standard and routine calculation which we omit. Symmetry when $(\underline{F}_a^{res}, E^{res})$ is transient follows from

$$\int dx\, \varphi(x)\, P_t^o\psi(x) = \mathscr{E}^{res} \int_{\zeta^*}^{\zeta} ds\varphi(X_s)\, \psi(X_{s+t}) e^{-a(\varkappa;t+s)+a(\varkappa;s)}$$

upon applying time reversal. In the recurrent case it suffices to apply the same argument to the operators $e^{-t} P_t^o$. By an obvious probabilistic argument

(18.7) $$G_u^{res} \varphi = G_u^o \varphi + G_u^o \varkappa G_u^{res} \varphi$$

and by symmetry also

(18.7') $$G_u^{res} \varphi = G_u^o \varphi + G_u^{res} \varkappa G_u^o \varphi .$$

Again it follows from these equations that the associated Dirichlet space is $(\underline{F}, E)$ and the theorem is proved. ///

The proof of Theorem 18.2 is easily adapted to establish the following more general result.

Theorem 18.3. (i) Let $\nu(dx) = \theta(x)\varkappa(dx)$ with $0 \leq \theta(x) \leq 1$ and let $(\tilde{\underline{F}}, \tilde{E})$ be defined in the same way as $(\underline{F}^{res}, E^{res})$ except that $\varkappa(dx)$ is replaced by $\nu(dx)$. Then $(\tilde{\underline{F}}, \tilde{E})$ is a regular Dirichlet space on $(\underline{X}, dx)$ and for $\varphi \geq 0$ the functions $\tilde{P}_t \varphi$ and $\tilde{G}_u \varphi$ are the minimal nonnegative solutions of

(18.8) $$\tilde{P}_t \varphi(x) = P_t \varphi(x) + \mathscr{E}_x I(t \leq \zeta ; X_{\zeta-0} \in \underline{X}) \theta(X_{\zeta-0}) \tilde{P}_{t-s} \varphi(X_{\zeta-0})$$

(18.8') $$\tilde{G}_u \varphi(x) = G_u \varphi(x) + \mathscr{E}_x e^{-u\zeta} I(X_{\zeta-0} \in \underline{X}) \theta(X_{\zeta-0}) \tilde{G}_u \varphi(X_{\zeta-0}) .$$

(ii) Let $\nu(dx)$ be a Radon measure charging no polar set and let $(\tilde{\underline{F}}, E)$ be defined by

$$\tilde{\underline{F}} = \underline{F} \cap L^2(\nu)$$

$$\tilde{E}(f,g) = E(f,g) + \int \nu(dx) f(x) g(x) .$$

Then $(\tilde{\underline{F}}, \tilde{E})$ is a regular Dirichlet space on $(\underline{X}, dx)$ and

18.5

$$\tilde{P}_t \varphi(x) = \mathcal{E}_x e^{-a(\nu;t)} \varphi(X_t) . \quad ///$$

Remark. Theorem A.1. in the appendix of [45] is valid only with some additional condition such as regularity. With this understanding, Theorem 18.3-(ii) is a refinement. ///

19. Equivalent Regular Representations

Let $(\underline{F}^o, E^o)$ be a general Dirichlet space on $L^2(\underline{X}^o, dx^o)$ such that $\underline{F}^o$ is dense in $L^2(\underline{X}^o, dx^o)$ and let $(\underline{F}, E)$ be a regular Dirichlet space on $L^2(\underline{X}, dx)$ following Fukushima [21] we make the

19.1. Definition. $(\underline{F}, E)$ is a regular representation of $(\underline{F}^o, E^o)$ if there exists a surjective isometry j^o from $L^2(\underline{\underline{X}}^o, dx^o)$ to $L^2(\underline{\underline{X}}, dx)$ such that $j^o Tf = Tj^o f$ whenever T is a mapping from R to R satisfying (1.12) and such that $j^o f$ belongs to $\underline{\underline{F}}$ if and only if f belongs to $\underline{\underline{F}}_0$ and then $E(j^o f, j^o f) = E(f,f)$. ///

Often in later sections $\underline{X}^o$ will be naturally imbedded in $\underline{X}$ in such a way that $dx^o = dx$ and j^o is the identity. In this case we say that $\underline{\underline{X}}$ is a regularizing space for $\underline{\underline{F}}^o$.

It follows from the construction in Section 2 that $(\underline{\underline{F}}^o, E^o)$ always has at least one regular representation and in general has more than one. Also it is easy to see that the construction in Section 2 gives the most general regular representation of $(\underline{\underline{F}}^o, E^o)$. We will verify in this section that the reflected and enveloping spaces are independent of possible choices of a regular representation. This will permit us to impose certain regularity conditions on a regular Dirichlet space $(\underline{\underline{F}}, E)$ without loss of generality, at least from the point of view of structure theory. Our main tool is the concept of a "quasi-homeomorphism" introduced by Fukushima in [22]. Our treatment of quasi-homeomorphisms will differ only slightly from [22].

19.2. Definition. Two regular Dirichlet spaces are equivalent if each represents the other. ///

Evidently two regular representations of the same Dirichlet space are equivalent. Therefore it suffices to consider two equivalent regular Dirichlet spaces $(\underline{\underline{F}}, E)$ and $(\underline{\underline{F}}', E')$ on $L^2(\underline{\underline{X}}, dx)$ and $L^2(\underline{\underline{X}}', dx')$ respectively. Let $j: L^2(\underline{\underline{X}}, dx) \to L^2(\underline{\underline{X}}', dx')$ implement the representation of $(\underline{\underline{F}}, E)$ by $(\underline{\underline{F}}', E')$. We first impose

19.3. Auxiliary Condition. $j(\underline{\underline{F}} \cap C_{com}(\underline{\underline{X}}))$ is contained in $\underline{\underline{F}}' \cap C_{com}(\underline{\underline{X}}'))$.///

It is easy to see that there exists a collection $\underline{\underline{B}}_0$ of functions in $\underline{\underline{F}}$ which satisfy the conditions 2.2.1 through 2.2.4 such that $\underline{\underline{X}}'$ can be identified with the maximal ideal space of the uniform closure $\underline{\underline{B}}$ and then j is identical with the isometry in Section 2. Moreover it follows from 19.3 that $C_{com}(\underline{\underline{X}})$ is contained in B and so there exists a continuous surjection $q': \underline{\underline{X}} \cup \{\partial'\} \to \underline{\underline{X}} \cup \{\partial\}$ such that $f(q'x') = jf(x')$ for f in $C_{com}(\underline{\underline{X}})$ and such that $q'\partial' = \partial$. Of course ∂' is the "dead point" adjoined to $\underline{\underline{X}}'$ in the same way that ∂ is adjoined to $\underline{\underline{X}}$. Since each f in $\underline{\underline{B}}_0$ can be approximated quasi-uniformly by functions in $C_{com}(\underline{\underline{X}})$ and since $\underline{\underline{B}}_0$ is countable, there exists an increasing sequence of closed subsets $\underline{\underline{X}}_n$ of $\underline{\underline{X}}$ such that $Cap_1(\underline{\underline{X}}-\underline{\underline{X}}_n) \to 0$ and such that each f has a version in $C_0(\underline{\underline{X}}_n)$. In order to guarantee that the versions in $C_0(\underline{\underline{X}}_n)$ are unique we assume further that each $\underline{\underline{X}}_n$ is dx regular in the sense of [22]. This means that $U_x \cap \underline{\underline{X}}_n$ is not dx null whenever x is in $\underline{\underline{X}}_n$ and U_x is a neighborhood of x. This can always be accomplished by replacing $\underline{\underline{X}}_n$ by $\tilde{\underline{\underline{X}}}_n$, the set of x in $\underline{\underline{X}}_n$ with the above property. The point is that $\tilde{X}_n$ is closed and has the same dx measure as $\underline{\underline{X}}_n$ and therefore their complements have the same capacity. Let $\underline{\underline{X}}_0 = \cup \underline{\underline{X}}_n$. Then $\underline{\underline{X}}_0$ is an F_σ set and $\underline{\underline{X}} - X_0$ is polar. There is a unique map $q: \underline{\underline{X}}_0 \cup \{\partial\} \to \underline{\underline{X}}' \cup \{\partial'\}$ such that $jf(qx) = f(x)$ for x in $\underline{\underline{X}}_0$ and f in $\underline{\underline{B}}_0$ and such that $q\partial = \partial'$. Clearly the composition

$q' \cdot q$ is the identity on $\underline{\underline{X}}_0 \cup \{\partial\}$ and in particular q is injective. Also for each n the mapping q is continuous when restricted to $\underline{\underline{X}}_n \cup \{\partial\}$. Thus $\underline{\underline{X}}'_n = q(\underline{\underline{X}}_n)$ is closed and q is a homeomorphism from $\underline{\underline{X}}_n$ to $\underline{\underline{X}}'_n$. We claim that $\underline{\underline{X}}' - \underline{\underline{X}}'_0$ is polar for $\underline{\underline{F}}'$ where $\underline{\underline{X}}'_0 = \bigcup \underline{\underline{X}}'_n$. If this were not the case then it would follow from Corollary 3.12 that there exists a nontrivial Radon measure μ' on $\underline{\underline{X}}'$ having finite E'_1 energy such that $\mu'(\underline{\underline{X}}'_0) = 0$. But then there exists a Radon measure μ on $\underline{\underline{X}}_0$ having finite energy such that

$$\int \mu(dx) f(x) = \int \mu'(dx')\, jf(x')$$

for f in $\underline{\underline{F}}$. Also there exists a Borel measure μ'' concentrated on $\underline{\underline{X}}'_0$ such that

$$\int \mu(dx)\, \varphi(qx) = \int \mu''(dx')\varphi(x')$$

for Borel $\varphi \geq 0$ on $\underline{\underline{X}}'_0$. But then clearly

$$\int \mu'(dx') f'(x') = \int \mu''(dx') f'(x')$$

for f' in $\underline{\underline{F}}' \cap C_{com}(\underline{\underline{X}}')$ which is impossible. Thus $\underline{\underline{X}}' - \underline{\underline{X}}'_0$ is indeed polar and it follows directly that actually $\mathrm{Cap}'_1(\underline{\underline{X}}' - \underline{\underline{X}}'_n) = \mathrm{Cap}_1(\underline{\underline{X}} - \underline{\underline{X}}_n)$.

19.4. Definition. Let $(\underline{\underline{F}}, E)$ and $(\underline{\underline{F}}', E')$ be equivalent regular Dirichlet spaces and let $j\colon L^2(X, dx) \to L^2(X', dx')$ implement the equivalence. A *quasi-homeomorphism* connecting $(\underline{\underline{F}}, E)$ to $(\underline{\underline{F}}', E')$ is a mapping $q\colon \underline{\underline{X}} \to \underline{\underline{X}}'$ which is defined quasi-everywhere on $\underline{\underline{X}}$ and satisfies the following conditions.

19.4.1. For each $\varepsilon > 0$ there exists a closed subset F of $\underline{\underline{X}}$ such that q restricted to F is a homeomorphism onto $F' = q(F)$ which is closed in $\underline{\underline{X}}'$ and $\mathrm{Cap}_1(\underline{\underline{X}} - F)$, $\mathrm{Cap}'_1(\underline{\underline{X}}' - F') < \varepsilon$.

19.4.2. q is measure preserving.

19.4.3. $jf \circ q = f$ for f in $L^2(\underline{\underline{X}}, dx)$. ///

We have shown above that if $(\underline{\underline{F}}',E')$ and $(\underline{\underline{F}},E)$ are equivalent and if the auxiliary condition 19.3 is satisfied then $(\underline{\underline{F}},E)$ is connected to $(\underline{\underline{F}}',E')$ by a quasi-homeomorphism. Clearly the latter property is symmetric and transitive and it follows easily that condition 19.3 can be dropped. We have proved

Theorem 19.1. *If $(\underline{\underline{F}},E)$ and $(\underline{\underline{F}}',E')$ are equivalent regular Dirichlet spaces then $(\underline{\underline{F}},E)$ is connected to $(\underline{\underline{F}}',E')$ by a quasi-homeomorphism.* ///

Now consider regular Dirichlet spaces $(\underline{\underline{F}},E)$ and $(\underline{\underline{F}}',E')$ as above which are connected by a quasi-homeomorphism q. Since

$$E(f,g) = -\int J(dy,dz)f(y)g(z)$$

whenever f,g in $\underline{\underline{F}} \cap C_{com}(\underline{\underline{X}})$ have disjoint supports and since in the transient case

$$\int \varkappa(dx)f(x) = \operatorname{Lim}_{t \downarrow 0} (1/t) \int dx \, \{1-P_t 1(x)\} \, f(x)$$

for f in $\underline{\underline{F}} \cap C_{com}(\underline{\underline{X}})$ (see Lemma 11.1), the Lévy kernel and killing measure are connected by

$$(19.1) \qquad \int \varkappa'(dx')f'(x') = \int \varkappa(dx)\, f'(qx)$$

$$(19.2) \qquad \iint J'(dx',dy')\, F(x',y') = \iint J(dx,dy)F(qx,qy).$$

Also it follows with the help of (11.9) that

$$(19.3) \qquad \int <Af'>(dx')\, \varphi'(x') = \int <A(f'\cdot q)>(dx)\varphi'(qx)$$

for f' in $\underline{\underline{F}}'$ and $\varphi' \geq 0$ and therefore

$$(19.4) \qquad \int <A_c f'>(dx')\, \varphi(x') = \int <A_c(f'\cdot q)>(dx)\, \varphi'(qx).$$

Now consider bounded f' in $\underline{\underline{F}}'^{ref}$ or $\underline{\underline{F}}'^{env}$. To show that $f = f'\cdot q$ is in $\underline{\underline{F}}^{ref}$, respectively $\underline{\underline{F}}^{env}$, it suffices by the above relations to show that f is in

$\underline{\underline{F}}_{loc}$. To see this fix U open with compact closure in $\underline{\underline{X}}$, choose g in $\underline{\underline{F}}$ such that g = 1 quasi-everywhere on U and let g' in $\underline{\underline{F}}'$ be such that g = g'· q. But then g'f' is in $\underline{\underline{F}}'$ and so gf=(g'f') · q is in $\underline{\underline{F}}$. This proves

Theorem 19.2. *Let* $(\underline{\underline{F}},E)$ *and* $(\underline{\underline{F}}',E')$ *be equivalent regular Dirichlet spaces and let* q *be a quasi-homeomorphism connecting* $(\underline{\underline{F}},E)$ *to* $(\underline{\underline{F}}',E')$.

(i) *The killing measures, Levy kernel and Dirichlet measures are related by* (19.1) *through* (19.4).

(ii) f' *belongs to* $\underline{\underline{F}}'^{ref}$ *if and only if* f = f' · q *belongs to* $\underline{\underline{F}}^{ref}$ *and then* $E'(f',f') = E(f,f)$.

(iii) f' *belongs to* $\underline{\underline{F}}'^{env}$ *if and only if* f = f' ·q *belongs to* $\underline{\underline{F}}^{env}$ *and then* $E'^{res}(f',f') = E^{res}(f,f)$. ///

Remark. In general the analogues of (ii) and (iii) are not valid for $\underline{\underline{F}}_{loc}$. ///

20. Second Structure Theorem

We begin with

Theorem 20.1. Let $(\underline{\underline{F}}, E)$ be a regular Dirichlet space on $L^2(\underline{\underline{X}}, dx)$ and let $(\underline{\underline{F}}^{\sim}, E^{\sim})$ be a second Dirichlet space on $L^2(\underline{\underline{X}}, dx)$, not necessarily regular. Then the following statements are equivalent.

(i) $\underline{\underline{F}}_b$ is an ideal in $\underline{\underline{F}}_b^{\sim}$ and $E(f,g) = E^{\sim}(f,g)$ for f,g in $\underline{\underline{F}}$.

(ii) $(\underline{\underline{F}}^{\sim}, E^{\sim})$ has a regularizing space $\underline{\underline{X}}^{\sim}$ such that $\underline{\underline{X}}$ is intrinsically open in $\underline{\underline{X}}^{\sim}$ and $(\underline{\underline{F}}, E)$ is the absorbed space for $\underline{\underline{X}}$. Moreover after possibly replacing $(\underline{\underline{F}}, E)$ by an equivalent representation we can choose $\underline{\underline{X}}^{\sim}$ so that $\underline{\underline{X}}$ is actually open. ///

Proof. It follows directly from Theorem 7.3-(iii) that (ii) implies (i). For the converse assume first that it is possible to choose $\underline{\underline{B}}_o^{\sim}$ satisfying 2.2.1 through 2.2.4 for $\underline{\underline{F}}^{\sim}$ and in addition

20.1.1. $\underline{\underline{B}}_o^{\sim} \cap C_{com}(\underline{\underline{X}})$ is uniformly dense in $C_{com}(\underline{\underline{X}})$.

20.1.2. Every f in $\underline{\underline{B}}_o^{\sim}$ is continuous on $\underline{\underline{X}}$.

Then the maximal ideal space $\underline{\underline{X}}^{\sim}$ is a regularizing space for $\underline{\underline{F}}^{\sim}$ and $\underline{\underline{X}}$ is naturally imbedded as an open subset. The ideal property of $\underline{\underline{F}}_b$ guarantees that $\underline{\underline{F}}^{\sim} \cap C_{com}(\underline{\underline{X}}) = \underline{\underline{F}} \cap C_{com}(\underline{\underline{X}})$ and (ii) follows with the help of Lemma 10.2-(i). To handle the general case it suffices to observe that the quasi-homeomorphic image of an open set is intrinsically open and that $\underline{\underline{B}}_o^{\sim}$ satisfying 20.1.1 and 20.1.2 can always be found if we are willing to first replace $(\underline{\underline{F}}, E)$ by an equivalent regular representation. ///

20.2. Terminology. Any Dirichlet space $(\underline{F}^{\sim},E^{\sim})$ satisfying the two equivalent conditions of Theorem 20.1 will be called an extension of $(\underline{F},E)$. ///

The Second Structure Theorem will be applicable to all Dirichlet spaces $(F^{\sim},E^{\sim})$ which are extensions of $(\underline{F},E)$. The examples at the end of Section 17 show that in general this class of Dirichlet spaces is wider than the class with generator $A^{\sim}$ contained in the local generator $\mathcal{O}$.

Now fix $(\underline{F}^{\sim},E^{\sim})$ a Dirichlet space on $L^2(\underline{X},dx)$ which is an extension of $(\underline{F},E)$. Replace $(\underline{F},E)$ by an equivalent representation if necessary and choose a regularizing space $\underline{X}^{\sim}$ for $(\underline{F}^{\sim},E^{\sim})$ such that $\underline{X}$ is open in $\underline{X}^{\sim}$ and $(\underline{F},E)$ is the absorbed space for $\underline{X}$. Put

$$\Delta = \underline{X}^{\sim} - \underline{X}$$

and adjoin a dead point $\partial^{\sim}$ to $\underline{X}^{\sim}$ according to the usual conventions. Clearly

$$(20.1) \qquad J^{\sim}(dy,dz) = J(dy,dz) \qquad \text{on } \underline{X}\times\underline{X}$$

$$(20.2) \qquad J^{\sim}(dy,\Delta) + \kappa^{\sim}(dy) = \kappa(dy) \qquad \text{on } \underline{X}.$$

Also $\underline{F}^{\sim}$ is contained in $\underline{F}_{loc}$ and

$$(20.3) \qquad \langle A^{\sim}_c f\rangle(dy) = \langle A_c f\rangle(dy) \qquad \text{on } \underline{X}.$$

Let $\mu(x,d\cdot)$, $x\in\underline{X}$ be subprobabilities on Δ satisfying the usual regularity conditions (see the lines following (10.10))and such that

$$(20.4) \qquad J^{\sim}(dx,dy) = \kappa(dx)\mu(x,dy) \qquad \text{on } \underline{X}\times\Delta\ .$$

This makes sense by (20.2). Also let

$$\mu(x) = \mu(x,\Delta);\quad r(x) = 1-\mu(x)$$

and then it is clear that

(20.5) $$\varkappa^{\sim}(dx) = r(x)\,\varkappa(dx) \qquad \text{on } \underline{\underline{X}}.$$

<u>20.3. Notation.</u> The symbols $X^{\sim}_{\zeta-0}$ and $X^{\sim}_{\zeta*}$ are used in place of $X_{\zeta-0}$ and $X_{\zeta*}$ when they are interpreted as points in $\underline{\underline{X}}^{\sim} \cup \{\partial^{\sim}\}$. For φ defined on Δ

$$\varphi^{\sim}(X^{\sim}_{\zeta-0}) = I(X^{\sim}_{\zeta-0} \in \Delta)\,\varphi(X^{\sim}_{\zeta-0}) + I(X^{\sim}_{\zeta-0} \in \underline{\underline{X}}) \int \mu(X^{\sim}_{\zeta-0}, dy)\,\varphi(y)$$

$$\varphi^{\sim}(X^{\sim}_{\zeta*}) = I(X^{\sim}_{\zeta*} \in \Delta)\,\varphi(X^{\sim}_{\zeta*}) + I(X^{\sim}_{\zeta*} \in \underline{\underline{X}}) \int \mu(X^{\sim}_{\zeta*}, dy)\,\varphi(y). \;///$$

Hitting operators for Δ will be denoted simply by $H^{\sim}$ and $H^{\sim}_{u}$. Thus

$$H^{\sim}\varphi(x) = \mathcal{E}^{\sim}_{x}\, I(\sigma(\Delta) < +\infty)\,\varphi(X_{\sigma(\Delta)})$$

$$H^{\sim}_{u}\varphi(x) = \mathcal{E}^{\sim}_{x}\, e^{-u\sigma(\Delta)}\varphi(X_{\sigma(\Delta)}).$$

It is easy to check that for $x \in \underline{\underline{X}}$ also

(20.6) $$H^{\sim}\varphi(x) = \mathcal{E}_{x}\,\varphi^{\sim}(X^{\sim}_{\zeta-0}).$$

$$H^{\sim}_{u}\varphi(x) = \mathcal{E}_{x}\, e^{-u\zeta}\,\varphi^{\sim}(X^{\sim}_{\zeta-0})$$

Let ν be the $E^{\sim}_{1}$ balayage of dx onto Δ :

(20.7) $$\nu(\Gamma) = \int dx\; H^{\sim}_{1} 1_{\Gamma}(x).$$

Let

(20.8) $$\underline{\underline{H}}^{\sim} = \gamma\underline{\underline{F}}^{\sim}$$

where γ is restriction to Δ and for $\varphi \in H^{\sim}$ let

(20.9) $$Q^{\sim}(\varphi,\varphi) = E^{\sim}(H^{\sim}\varphi, H^{\sim}\varphi)\;;\quad Q^{\sim}_{(u)}(\varphi,\varphi) = E^{\sim}_{u}(H^{\sim}_{u}\varphi,\, H^{\sim}_{u}\varphi).$$

Then $(\underline{\underline{H}}^{\sim}, Q^{\sim})$ is the time changed Dirichlet space of Section 8 with $(\underline{\underline{F}}^{\sim}, E^{\sim})$ playing the role of (F, E). In the notation of that section the operator $R^{\nu}_{(1)}$ is bounded on L^{∞} and therefore, by symmetry, on $L^2(\nu)$. From this it follows that the Dirichlet norm $Q^{\sim}_{(1)}$ dominates a multiple of the standard inner product on $L^2(\nu)$ and therefore

(20.10) $$\underline{\underline{H}}^{\sim} \subset L^2(\Delta, \nu) .$$

Let $\pi^{\sim}_u$, $u > 0$ be the adjoint operators from $L^2(\underline{\underline{X}}, dx)$ to $L^2(\Delta, \nu)$ determined by

(20.11) $$\int dx\, f(x)\, H^{\sim}_u \varphi(x) = \int \nu(dy)\, \pi^{\sim}_u f(y) \varphi(y) .$$

Clearly $H^{\sim}_1$ is bounded from $L^2(\nu)$ to $L^2(dx)$ and from the resolvent like identity

(20.12) $$H^{\sim}_u = H^{\sim}_v + (v-u)\, G_u H^{\sim}_v$$

and its dual

(20.12') $$\pi^{\sim}_u = \pi^{\sim}_v + (v-u)\, \pi^{\sim}_v H^{\sim}_u$$

follows the boundedness of $H^{\sim}_u$, $u > 0$ and also of $\pi^{\sim}_u$, $u > 0$ from $L^2(dx)$ to $L^2(\nu)$.

For $0 \leq u < v$ we introduce the associated <u>Feller density</u> $U^{\sim}_{u,v}(y,z)$. This is jointly measurable on $\Delta \times \Delta$ and is specified up to $\nu \times \nu$ equivalence by

(20.13) $$\int \nu(dy) \int \nu(dz)\, U^{\sim}_{u,v}(y,z)\, \varphi(y)\, \psi(z)$$

$$= (v-u) \int dx\, H^{\sim}_u \varphi(x)\, H^{\sim}_v \psi(z)$$

$$= (v-u) \int \nu(dy) \varphi(y)\, \pi^{\sim}_u H^{\sim}_v \psi(y) .$$

The right side of (20.13)

$$= (v-u)\ \mathscr{E}\int_{\zeta^*}^{\zeta} dt\ I(\zeta<\infty)\ e^{-u(\zeta-t)}\varphi^{\sim}(X^{\sim}_{\zeta-0})\ H^{\sim}_{v}\psi(X_t)$$

$$= (v-u)\ \mathscr{E}\int_{\zeta^*}^{\zeta} dt\ I(\zeta^*>-\infty)e^{-u(t-\zeta^*)}\varphi^{\sim}(X^{\sim}_{\zeta^*})\ H^{\sim}_{v}\psi(X_t)$$

$$= (v-u)\mathscr{E}\int_{\zeta^*}^{\zeta} dt\ I(-\infty<\zeta^*<\zeta<+\infty)e^{-u(t-\zeta^*)}e^{-v(\zeta-t)}\varphi^{\sim}(X^{\sim}_{\zeta^*})\ \psi^{\sim}(X^{\sim}_{\zeta-0})$$

and therefore also

(20.13') $$\int\nu\ (dy)\int\nu(dz)\ U^{\sim}_{u,v}(y,z)\varphi\ (y)\psi\ (z)$$

$$= \mathscr{E}\ I(-\infty<\zeta^*<\zeta<+\infty)\varphi^{\sim}(X^{\sim}_{\zeta^*})\psi^{\sim}(X^{\sim}_{\zeta-0})\{e^{-u(\zeta-\zeta^*)}-e^{-v(\zeta-\zeta^*)}\}.$$

It follows in particular that

(20.14) $$U^{\sim}_{u,w}(y,z) = U^{\sim}_{u,v}(y,z) + U^{\sim}_{v,w}(y,z) \qquad 0\leq u<v<w$$

and that $U^{\sim}_{u,\infty}$ is well defined either by (20.13') or by

(20.15) $$U^{\sim}_{u,\infty}(y,z) = \lim_{v\uparrow\infty} U^{\sim}_{u,v}(y,z).$$

We introduce also the special notations

(20.16) $$U^{\sim}_{u,v}(\varphi,\psi) = \int\nu(dy)\int\nu\ (dz)\ U^{\sim}_{u,v}(y,z)\varphi\ (y)\psi\ (z)$$

$$U^{\sim}_{u,v}<\varphi,\varphi> = \int\nu\ (dy)\int\nu\ (dz)\ U^{\sim}_{u,v}(y,z)\{\varphi\ (y)-\varphi(z)\}^2.$$

The relation

(20.17) $$U^{\sim}_{u,v}(1,\varphi^2) = \tfrac{1}{2}\ U^{\sim}_{u,v}<\varphi,\varphi> + U^{\sim}_{u,v}(\varphi,\varphi)$$

is easily established using symmetry of $U^{\sim}_{u,v}(y,z)$.

The <u>Naim kernel</u> is the measure $\Theta(dy,dz)$ on $\Delta\times\Delta$ determined by

$$(20.18)\qquad \Theta(\Gamma',\Gamma) = \mathcal{E}\, I(-\infty<\zeta^*<\zeta<+\infty;\ \tilde{X}_{\zeta^*}, \tilde{X}_{\zeta-0} \in \Delta)\, 1_{\Gamma'}(\tilde{X}_{\zeta^*})\, 1_{\Gamma}(\tilde{X}_{\zeta-0}).$$

The associated Naim kernel $\Theta^{\sim}(dy,dz)$ is defined on $\Delta \times \Delta$ by

$$(20.19)\qquad \iint \Theta^{\sim}(dy,dz)\varphi(y)\psi(z) = \mathcal{E}\,\tilde{\varphi}(\tilde{X}_{\zeta^*})\tilde{\psi}(\tilde{X}_{\zeta-0}).$$

It is easy to check that $\Theta^{\sim}$ and Θ are related by

$$(20.20)\qquad \Theta^{\sim}(\Gamma',\Gamma) = \Theta(\Gamma',\Gamma) + \int \varkappa(dx)\mu(x,\Gamma')\, H^{\sim} 1_{\Gamma}(x)$$

$$+ \int \varkappa(dx)\mu(x,\Gamma) H^{\sim} 1_{\Gamma'}(x)$$

$$+ \iint (\varkappa N \varkappa)(dx,dw)\mu(x,\Gamma)\mu(w,\Gamma').$$

Here $\varkappa N \varkappa$ is the measure on $\underline{\underline{X}} \times \underline{\underline{X}}$ determined by

$$(20.21)\qquad \iint (\varkappa N\varkappa)(dx,dw)\varphi(x)\psi(w)$$

$$= \int \varkappa(dx)\varphi(x)\, N(\psi\cdot\varkappa)(x)$$

$$= \int \varkappa(dw)\psi(w)\, N(\varphi\cdot\varkappa)(w).$$

The <u>associated excursion form</u> is

$$(20.22)\qquad N^{\sim}(\varphi,\varphi) = \tfrac{1}{2}\mathcal{E}\{\varphi^{\sim}(\tilde{X}_{\zeta^*}) - \varphi^{\sim}(\tilde{X}_{\zeta-0})\}^2 .$$

The <u>associated excursion space</u> $\underline{\underline{N}}^{\sim}$ is the set of φ in $L^2(\Delta,\nu)$ for which (20.22) converges. Clearly (20.22) is equivalent to

$$(20.23)\qquad N^{\sim}(\varphi,\varphi) = \tfrac{1}{2}\iint \Theta^{\sim}(dy,dz)\{\varphi(y)-\varphi(z)\}^2$$

$$+ \mathcal{E}\, I(\tilde{X}_{\zeta^*} = \tilde{\partial})\, \{\varphi^{\sim}(\tilde{X}_{\zeta-0})\}^2$$

$$+ \int \varkappa(dx) r(x)\, H^{\sim}\varphi^2(x).$$

It is important also that for $u>0$

(20.24) $$\tilde{N}(\varphi,\varphi) = \tfrac{1}{2}\, \tilde{U}_{0,\infty} <\varphi,\varphi> + \int \varkappa(dx) r(x)\, \tilde{H}\varphi^2(x)$$

$$+ u\int dx\, \tilde{p}(x)\, \tilde{H}_u \varphi^2(x)$$

$$+ \mathcal{E}\{1-e^{-u(\zeta-\zeta^*)}\}\, I(\tilde{X}_{\zeta^*} = \tilde{\partial})\{\tilde{\varphi}(\tilde{X}_{\zeta-0})\}^2$$

where

(20.25) $$\tilde{p}(x) = P_x[\tilde{X}_{\zeta^*} = \tilde{\partial}].$$

This follows on the one hand from the relation

(20.26) $$\tilde{U}_{0,\infty}(y,z)\nu(dy)\nu(dz) = \tilde{\Theta}(dy,dz)$$

which is an easy consequence of (20.13') and on the other hand from the calculation

(20.27) $$u\int dx\, \tilde{p}(x)\, \tilde{H}_u \varphi^2(x)$$

$$= \mathcal{E}\int_{\zeta^*}^{\zeta} udt\, p(X_t)e^{-u(\zeta-t)}\{\tilde{\varphi}(\tilde{X}_{\zeta-0})\}^2$$

$$= \mathcal{E}\int_{\zeta}^{\zeta} udt\, p(X_t)e^{-u(t-\zeta^*)}\{\tilde{\varphi}(\tilde{X}_{\zeta^*})\}^2$$

$$= \mathcal{E}\{\tilde{\varphi}(\tilde{X}_{\zeta^*})\}^2\int_{\zeta^*}^{\zeta} udt\, e^{-u(t-\zeta^*)} I(\tilde{X}_{\zeta-0} = \tilde{\partial})$$

$$= \mathcal{E}\{\tilde{\varphi}(\tilde{X}_{\zeta^*})\}^2 e^{-u(\zeta-\zeta^*)}$$

$$= \mathcal{E}\{\tilde{\varphi}(\tilde{X}_{\zeta-0})\}^2 e^{-u(\zeta-\zeta^*)}.$$

(In the special case of Brownian motion, (20.26) is the relation established by M. Fukushima in [18].) Now we are ready for

Theorem 20.2. (Second Structure Theorem) Let $(F^{\sim}, E^{\sim})$ be an extension of (F, E) and assume that $(F^{\sim}, E^{\sim})$ has a regularizing space $X^{\sim}$ such that X is open in $X^{\sim}$ and (F, E) is the absorbed space for X.

(i) There is no nontrivial $\varphi \geq 0$ in $L^2(\Delta, \nu)$ such that $U^{\sim}_{0,\infty}(1, \varphi^2) < +\infty$.

(ii) The pair $(H^{\sim}, Q^{\sim})$ is a Dirichlet space on $L^2(\Delta, \nu)$.

(iii) $H^{\sim}$ is contained in $N^{\sim}$ and $Q^{\sim} - N^{\sim}$ is contractive on $H^{\sim}$.

(iv) For f in $F^{\sim}$ the difference $f - H^{\sim}\gamma f$ is in $F_{(e)}$ and

$$(20.28) \qquad E^{\sim}(f, f) = Q^{\sim}(\gamma f, \gamma f) + E(f - H^{\sim}\gamma f, f - H^{\sim}\gamma f).$$

Moreover if φ is in $H^{\sim}$ then $H^{\sim}\varphi$ is in $F^{\sim}_{(e)}$ and of course $E^{\sim}(H^{\sim}\varphi, H^{\sim}\varphi) = Q^{\sim}(\varphi, \varphi)$.

(v) For $u > 0$ the pair $(H^{\sim}, Q^{\sim}_{(u)})$ is a Dirichlet space on $L^2(\Delta, \nu)$. For f in $F^{\sim}$ the difference $f - H^{\sim}_u \gamma f$ is in F and

$$(20.29) \qquad E^{\sim}_u(f, f) = Q^{\sim}_{(u)}(\gamma f, \gamma f) + E_u(f - H^{\sim}_u \gamma f, f - H^{\sim}_u \gamma f).$$

Also if φ is in $H^{\sim}$ then $H^{\sim}_u \varphi$ is in $F^{\sim}$ and $E^{\sim}_u(H^{\sim}_u\varphi, H^{\sim}_u\varphi) = Q^{\sim}_{(u)}(\varphi, \varphi)$.

(vi) The associated resolvent operators $G^{\sim}_u, u > 0$ can be represented

$$(20.30) \qquad G^{\sim}_u = G_u + H^{\sim}_u R^{\sim}_{(u)} \pi^{\sim}_u$$

where $R^{\sim}_{(u)}$ is the Green's operator for $(H^{\sim}, Q^{\sim}_{(u)})$.

Conversely let Δ be a separable locally compact Hausdorff space and let $X^{\sim} = X \cup \Delta$. Adjoin a dead point $\partial^{\sim}$ to $X^{\sim}$ and assume that $X^{\sim} \cup \{\partial^{\sim}\}$ is topologized in such a way that X is topologically imbedded as an open set and for quasi-every x in X the limit $X_{\zeta-0}$ exists as a point in $X^{\sim} \cup \{\partial^{\sim}\}$ on the set $[X_{\zeta-0} = \partial]$ except for a P_x null set. Let $\mu(x, d\cdot)$, $x \in X$ be a family of subprobabilities on Δ satisfying the usual regularity conditions and let the

Notation 20.3 be in effect. Let $H^\sim$, $H^\sim_u$, ν, $\pi^\sim_u$, $U^\sim_{u,v}$, $N^\sim$, $\underline{\underline{N}}$ be defined by (20.6), (20.7), (20.11), (20.13) and (20.22). Let $(\underline{\underline{H}}^\sim, Q^\sim)$ be a Dirichlet space on $L^2(\Delta, \nu)$ such that $\underline{\underline{H}}^\sim$ is contained in $\underline{\underline{N}}^\sim$ and $Q^\sim - N^\sim$ is contractive on $\underline{\underline{H}}^\sim$. For $u > 0$ define

(20.31) $$Q^\sim_{(u)}(\varphi,\varphi) = Q^\sim(\varphi,\varphi) + U^\sim_{0,u}(\varphi,\varphi).$$

Then $(\underline{\underline{H}}^\sim, Q^\sim_{(u)})$ is also a Dirichlet space on $L^2(\Delta, \nu)$ with Green's operator $R^\sim_{(u)}$ bounded on $L^2(\Delta, \nu)$. The operators $G^\sim_u$ defined by (20.30) form a symmetric submarkovian resolvent on $L^2(X, dx)$. If condition (i) above is satisfied then the associated Dirichlet space $(\underline{\underline{F}}^\sim, E^\sim)$ is an extension of $(\underline{\underline{F}}, E)$ and there is a unique map $\gamma : \underline{\underline{F}}^\sim \to \underline{\underline{H}}^\sim$ satisfying (iv) and (v). ///

Remark. If condition (i) fails then it may turn out that $(\underline{\underline{F}}^\sim, E^\sim)$ is not an extension of $(\underline{\underline{F}}, E)$ even though $P^\sim_t$ dominates P_t. Such Dirichlet spaces are treated in Section 21. It is clear that the topological hypotheses for the converse can be relaxed somewhat but we doubt this has much significance. ///

Proof. For (i) note that for $v > 1$

(20.32) $$U^\sim_{1,v}(\varphi,\varphi) = (v-1)\int dx\, H^\sim_1\varphi(x)\{1-(v-1)G_v\}\, H^\sim_1\varphi(x).$$

Thus if $U^\sim_{0,\infty}(\varphi,\varphi) < +\infty$ then $H^\sim_1\varphi$ is in $\underline{\underline{F}}$. But this is impossible for nontrivial φ in $L^2(\Delta, \nu)$ by Theorem 7.3-(i). Conclusions (ii), (iv) and (v) follow from the results on random time change in Section 8. Conclusion (iii) will follow if we establish

(20.33) $$N^\sim(\varphi,\varphi) = \tfrac{1}{2}\iint_{\underline{\underline{X}}\times\underline{\underline{X}}} J(dx,dw)\,\{H^\sim\varphi(x) - H^\sim\varphi(w)\}^2 + \iint_{\underline{\underline{X}}\times\Delta} J^\sim(dx,dy)\,\{H^\sim\varphi(x) - \varphi(y)\}^2$$

$$+\int_{\underline{\underline{X}}} \langle A_c H^{\sim}\varphi\rangle (dx)$$

$$+\int_{\underline{\underline{X}}} \kappa (dx) r(x) \{ H^{\sim}\varphi (x) \}^2$$

since then

$$Q^{\sim}(\varphi,\varphi) = N^{\sim}(\varphi,\varphi) + \tfrac{1}{2}\int\int_{\Delta\times\Delta} J^{\sim}(dy,dz)\{\varphi(y)-\varphi(z)\}^2 + \int_{\Delta} \langle A^{\sim}_c\varphi\rangle (dy) + \int_{\Delta} \kappa^{\sim}(dy)\varphi^2(y) \tag{20.34}$$

and in particular the difference $Q^{\sim} - N^{\sim}$ is contractive on $\underline{H}^{\sim}$ in the sense of (15.14). In verifying (20.33) we consider only the case when $(\underline{\underline{F}}^{\sim}, E^{\sim})$ is transient. The necessary modifications for the recurrent case are obvious from Section 16. We use the notation of Section 13 with $\underline{\underline{X}}$ playing the role of D. By (13.7') the right side of (20.33)

$$= \tfrac{1}{2}\mathcal{E}^{\sim}\Sigma_i I(r(i) < \zeta)\{\varphi(X_{r(i)}) - \varphi(X_{e(i)-0})\}^2$$

$$+ \tfrac{1}{2}\mathcal{E}^{\sim}\Sigma_i I(r(i) = \zeta)\varphi^2(X_{e(i)-0})$$

$$+ \tfrac{1}{2}\int_{\underline{\underline{X}}} \kappa (dx) r(x) H^{\sim}\varphi^2(x)$$

$$+ \tfrac{1}{2}\mathcal{E}^{\sim} I(X_{\zeta *} = \partial^{\sim};\ \zeta^* < \sigma(\Delta) < \zeta)\varphi^2(X_{\sigma(\Delta)}).$$

$$= \tfrac{1}{2}\mathcal{E}^{\sim}(\Sigma_i I(r(i) < \zeta)\{\varphi(X_{r(i)}) - \varphi(X_{e(i)-0})\}^2$$

$$+ \int_{\underline{\underline{X}}} \kappa (dx) r(x) H^{\sim}\varphi^2(x)$$

$$+ \mathcal{E}^{\sim} I(X_{\zeta *} = \partial^{\sim};\ \zeta^* < \sigma(\Delta) < \zeta)\varphi^2(X_{\sigma(\Delta)})$$

and so (20.33) will follow once we establish

$$(20.35)\qquad \mathcal{E}^{\sim} I(X_{\zeta*} = \partial^{\sim};\ \zeta* < \sigma(\Delta) < \zeta)\ \varphi^2(X_{\sigma(\Delta)})$$

$$= \mathcal{E}\ I(\tilde{X}_{\zeta*} = \partial)\ \{\varphi^2(\tilde{X}_{\zeta-0})\}^2$$

$$(20.36)\qquad \mathcal{E}^{\sim} \Sigma_i I(r(i) < \zeta)\{\varphi(X_{r(i)}) - \varphi(X_{e(i)-0})\}^2$$

$$= \tfrac{1}{2}\int\int \Theta^{\sim}(dy,dz)\ \{\varphi(y) - \varphi(z)\}^2 .$$

The left side of (20.35)

$$(20.37)\qquad = \operatorname{Lim}_{v\uparrow\infty} \mathcal{E}^{\sim} I(X_{\zeta*} = \partial^{\sim};\ \zeta* < \sigma(\Delta) < \zeta)\ \{1 - e^{-v[\sigma(\Delta)-\zeta*]}\}\ \varphi^2(X_{\sigma(\Delta)})$$

$$= \operatorname{Lim}_{v\uparrow\infty} \mathcal{E}^{\sim} I(X_{\zeta*} = \partial^{\sim};\ \zeta* < \sigma(\Delta) < \zeta) \int_{\zeta*}^{\sigma(\Delta)} dt\, v e^{-v[\sigma(\Delta)-t]}\ \varphi^2(X_{\sigma(\Delta)})$$

$$= \operatorname{Lim}_{v\uparrow\infty} \mathcal{E}^{\sim} I(X_{\zeta*} = \tilde{\partial}) \int_{\zeta*}^{\sigma(\Delta)} dt\, v\ \tilde{H}_v \varphi^2(X_t)$$

$$= \operatorname{Lim}_{v\uparrow\infty} \mathcal{E}^{\sim} \int_{\sigma*(\Delta)}^{\zeta} dt\ v\ \tilde{H}_v \varphi^2(X_t) I(X_{\zeta-0} = \tilde{\partial})$$

$$= \operatorname{Lim}_{v\uparrow\infty} \mathcal{E}^{\sim} \int_{\zeta*}^{\zeta} dt\ v\ \tilde{H}_v \varphi^2(X_t) I(\theta_t \sigma(\Delta) = +\infty;\ X_{\zeta-0} = \tilde{\partial})$$

$$= \operatorname{Lim}_{v\uparrow\infty} \mathcal{E}^{\sim} \int_{\zeta*}^{\zeta} dt\, v\, \tilde{H}_v \varphi^2(X_t) p^{\sim}(X_t)$$

$$= \operatorname{Lim}_{v\uparrow\infty} \int dx\, v \tilde{H}_v \varphi^2(x) p(x)$$

which by an analagous calculation equals the right side of (20.35). A similar argument works for (20.36). The left side

$$(20.38) \quad = \operatorname{Lim}_{v\uparrow\infty} \tilde{\delta}\, \Sigma_i I(r(i)<\zeta)\{1-e^{-v[r(i)-e(i)]}\}\{\varphi(X_{r(i)})-\varphi(X_{e(i)-0})\}^2$$

$$= \operatorname{Lim}_{v\uparrow\infty} \tilde{\delta}\, \Sigma_i I(r(i)<\zeta)\int_{e(i)}^{r(i)} dt\, v\, e^{-v[r(i)-t]}\{\varphi(X_{r(i)})-\varphi(X_{e(i)-0})\}^2$$

$$= \operatorname{Lim}_{v\uparrow\infty} \tilde{\delta}\, \Sigma_i \int_{e(i)}^{r(i)} dt\, v \int \tilde{H}_v(X_t, dz)\{\varphi(z)-\varphi(X_{e(i)-0})\}^2$$

$$= \operatorname{Lim}_{v\uparrow\infty} [\tilde{\delta}\, \Sigma_i I(r(i)<\zeta)\int_{e(i)}^{r(i)} dt\, v \int \tilde{H}_v(X_t, dz)\{\varphi(z)-\varphi(X_{r(i)})\}^2$$

$$+ \tilde{\delta}\int_{\zeta^*}^{\sigma(\Delta)} dt\, v \int \tilde{H}_v(X_t, dz)\{\varphi(z)-\varphi(X_{\sigma(\Delta)})\}^2]$$

$$= \operatorname{Lim}_{v\uparrow\infty} \tilde{\delta}\int_{\zeta^*}^{\zeta} dt\, 1_{\underline{\underline{X}}}(X_t)\, v \int \tilde{H}_v(X_t, dz) \int \tilde{H}_t(X_t, dy)\{\varphi(z)-\varphi(y)\}^2$$

$$= \operatorname{Lim}_{v\uparrow\infty} v \int dx \int \tilde{H}_v(x, dz) \int \tilde{H}(x, dy)\{\varphi(z)-\varphi(y)\}^2$$

and again by an analagous calculation this is also equal to the right side of (20.36).

We establish (vi) and the converse at the same time. Since

$$\tilde{Q}_{(u)}(\varphi,\varphi) = \tilde{Q}(\varphi,\varphi) - \tfrac{1}{2}\tilde{U}_{0,u}\langle\varphi,\varphi\rangle$$

$$+ \tilde{U}_{0,u}(1,\varphi^2)$$

is contractive by (20.24) and since

$$\tilde{U}_{0,1}(\varphi,\varphi) \le \tilde{U}_{0,1}(1,\varphi^2)$$

$$\le \int \nu(dy)\varphi^2(y)$$

and the forms $U^{\sim}_{0,u}(\varphi,\varphi)$, $u>0$ are all comparable, also each pair $(\underline{H}^{\sim}, Q^{\sim}_{(u)})$ is a Dirichlet space on $L^2(\Delta,\nu)$. Let $\{R^{\sim}_{(u)\alpha}, \alpha>0\}$ be the corresponding resolvent on $L^2(\Delta,\nu)$ and let $R^{\sim}_{(u)} = \mathrm{Lim}_{\alpha\downarrow 0} R^{\sim}_{(u)\alpha}$. By (20.24)

$$Q^{\sim}_{(u)}(\varphi,\varphi) - \int dx\, p^{\sim}(x) u\, H^{\sim}_u \varphi^2(x) - \int \varkappa(dx) r(x) H^{\sim}\varphi^2 - U_{0,u}(1,\varphi^2)$$

$$= Q^{\sim}(\varphi,\varphi) - \int dx\, p^{\sim}(x) u\, H^{\sim}_u \varphi^2(x) - \int \varkappa(dx) r(x) H^{\sim}\varphi^2(x) - \tfrac{1}{2} U_{0,u} < \varphi,\varphi>$$

is contractive and since

$$U_{0,u}(1,\varphi^2) + \int \varkappa(dx) r(x)\, H^{\sim}\varphi^2(x)$$

$$\geq u \int dx\, H^{\sim}1(x) H^{\sim}_u \varphi^2(x) + u \int \varkappa(dx) r(x) G\, H^{\sim}_u \varphi^2(x)$$

$$= u \int dx \{ H^{\sim}1(x) + N\, r\cdot\varkappa(x) \} H^{\sim}_u \varphi^2(x)$$

it follows by the proof of Theorem 12.1 that

$$uR^{\sim}_{(u)} \pi^{\sim}_u \{ p^{\sim} + H^{\sim}1 + Nr\cdot\varkappa \} \leq 1 \qquad [\text{a.e.}\, \nu^{\sim}]. \tag{20.39}$$

or, equivalently

$$uR^{\sim}_{(u)} \pi^{\sim}_u 1 \leq 1 \qquad [\text{a.e.}\, \nu^{\sim}]. \tag{20.39'}$$

(Notice that in establishing (20.39') for all $u>0$ we have used the contractivity of $Q^{\sim} - N^{\sim}$ in full strength.) Define $G^{\sim}_u, u>0$ by (20.30). With the help of (20.39') it is easy to see that the $G^{\sim}_u, u>0$ are symmetric and submarkovian. To establish the resolvent identity

$$G^{\sim}_u = G^{\sim}_v + (v-u)\, G^{\sim}_u G^{\sim}_v \tag{20.40}$$

note first that since for $0<u<v$

$$Q^{\sim}_{(v)}(\varphi,\varphi) = Q^{\sim}_{(u)}(\varphi,\varphi) + U^{\sim}_{u,v}(\varphi,\varphi)$$

a computation analagous to (12.5) establishes

$$R^{\sim}_{(u)} = R^{\sim}_{(v)} + (v-u)\, R^{\sim}_{(u)} \pi^{\sim}_{v} H^{\sim}_{u} R^{\sim}_{(v)} \tag{20.41}$$

$$= R^{\sim}_{(v)} + (v-u)\, R^{\sim}_{(v)} \pi^{\sim}_{v} H^{\sim}_{u} R^{\sim}_{(u)} \, .$$

Then (20.40) follows from the calculation

$$G^{\sim}_{u} - G^{\sim}_{v} = G_{u} - G_{v} + H^{\sim}_{u} R^{\sim}_{(u)} \pi^{\sim}_{u} - H^{\sim}_{v} R^{\sim}_{(v)} \pi^{\sim}_{v} \tag{20.42}$$

$$= (v-u)\, G_{u} G_{v} + H^{\sim}_{u} R^{\sim}_{(u)} \pi^{\sim}_{v} \{1 + (v-u)\, G_{u}\}$$

$$- \{1-(v-u)\, G_{u}\} H^{\sim}_{u} R^{\sim}_{(v)} \pi^{\sim}_{v}$$

$$= (v-u)\, G_{u} G_{v} + H^{\sim}_{u} \{ R^{\sim}_{(u)} - R^{\sim}_{(v)} \} \pi^{\sim}_{v}$$

$$+ (v-u)\, H^{\sim}_{u} R^{\sim}_{(u)} \pi^{\sim}_{v} G_{u} + (v-u)\, G_{v} H^{\sim}_{u} R^{\sim}_{(v)} \pi^{\sim}_{v}$$

$$= (v-u)\, G_{u} G_{v} + (v-u)\, H^{\sim}_{u} R^{\sim}_{(u)} \pi^{\sim}_{u} H^{\sim}_{v} R^{\sim}_{(v)} \pi^{\sim}_{v}$$

$$+ (v-u)\, H^{\sim}_{u} R^{\sim}_{(u)} \pi^{\sim}_{u} G_{v} + (v-u)\, G_{u} H^{\sim}_{v} R^{\sim}_{(v)} \pi^{\sim}_{v}$$

$$= (v-u)\, G^{\sim}_{u} G^{\sim}_{v} \, .$$

Now assume condition (i) and let $(\underline{\underline{F}}^{\sim}, E^{\sim})$ be the Dirichlet space on $L^2(\underline{\underline{X}}, dx)$ which corresponds to $G^{\sim}_{u}$, $u > 0$. First $G^{\sim}_{u} \geq G_{u}$ and so it follows directly from Lemma 1.1 that $\underline{\underline{F}}^{\sim}$ contains $\underline{\underline{F}}$ and that $E^{\sim}(f,f) \leq E(f,f)$ for f in $\underline{\underline{F}}$. Moreover if $f \in \underline{\underline{F}}^{\sim}$, $g \in \underline{\underline{F}}$ and $|f| \leq |g|$, then the estimate

(20.43) $$\int dx \{ f(x) - uG_u f(x) \} f(x)$$

$$\leq \tfrac{1}{2} \int dx \int \tilde{G}_u (x, dy) \{ f(x) - f(y) \}^2$$

$$+ \int dx \{ 1 - uG_u 1(x) \} g^2(x)$$

shows that f is in $\underline{\underline{F}}$ and it follows in particular that $\underline{\underline{F}}_b$ is an ideal in $\tilde{\underline{\underline{F}}}_b$. For g, g' in $L^2(\underline{X}, dx)$ and for $u > 0$ both $\tilde{H}_u \tilde{R}_{(u)} \tilde{\pi}_u g$ and $G_u g'$ are in $\underline{\underline{F}}^{\sim}$ and since

(20.44) $$v \int dx \{ \tilde{H}_u \tilde{R}_{(u)} \tilde{\pi}_u g(x) - v \tilde{G}_{u+v} \tilde{H}_u \tilde{R}_{(u)} \tilde{\pi}_u g(x) \} G_u g'(x)$$

$$= v \int dx \{ 1 - vG_{u+v} \} \tilde{H}_u \tilde{R}_{(u)} \tilde{\pi}_u g(x) G_u g'(x)$$

$$- v^2 \int dx \tilde{H}_{u+v} \tilde{R}_{(u+v)} \tilde{\pi}_{u+v} \tilde{H}_u \tilde{R}_{(u)} \tilde{\pi}_u g(x) G_u g'(x)$$

$$= v \int dx G_{u+v} \tilde{H}_u \tilde{R}_{(u)} \tilde{\pi}_{(u)} g(x) g'(x)$$

$$- v \int dx \tilde{H}_{u+v} \{ \tilde{R}_{(u)} - \tilde{R}_{(u+v)} \} \tilde{\pi}_u g(x) G_u g'(x)$$

$$= v \int dx \tilde{H}_u \tilde{R}_{(u+v)} \tilde{\pi}_u g(x) G_{u+v} g'(x)$$

(using (20.41; and also $G_{u+v} \tilde{H}_u = G_u \tilde{H}_{u+v}$), the relation

(20.45) $$\tilde{E}_u (\tilde{H}_u \tilde{R}_{(u)} \tilde{\pi}_u g \;, G_u g') = 0$$

will follow once we establish

(20.46) $$\operatorname{Lim}_{v \uparrow \infty} \tilde{R}_{(u+v)} = 0$$

in the strong operator topology on $L^2(\nu)$. But if (20.46) fails then there exists $\varphi \geq 0$ in $L^2(\Delta)$ such that $\psi = \operatorname{Lim}_{v \uparrow \infty} \tilde{R}_v \varphi$ is nontrivial and this contradicts (i) because of the estimate

20.16

$$\int \nu\,(dy)\varphi\,(y)\,R^{\sim}_{v}\varphi\,(y)$$

$$= Q^{\sim}_{(v)}(R^{\sim}_{v}\varphi\,,R^{\sim}_{v}\varphi\,)$$

$$\geq U^{\sim}_{0,v}(R^{\sim}_{v}\varphi\,,R^{\sim}_{v}\varphi\,) + U^{\sim}_{0,\infty}\langle R^{\sim}_{v}\varphi,\ R^{\sim}_{v}\varphi\rangle$$

$$\geq U^{\sim}_{0,v}\,(1,\psi^{2}).$$

This proves (20.45). For g, g' as above

$$E^{\sim}_{1}\,(G_{1}g, G_{1}g') = E^{\sim}_{1}(G^{\sim}_{1}g, G_{1}g')$$

$$= \int dx\, g(x)\; G_{1}g'(x)$$

$$= E_{1}(G_{1}g, G_{1}g')$$

and it follows directly that

(20.47) $\qquad E^{\sim}(f,f') = E(f,f')$

for f,f' in $\underline{\underline{F}}$. Thus $(\underline{\underline{F}}^{\sim}, E^{\sim})$ is indeed an extension of $(\underline{\underline{F}},E)$ and to complete the proof we need only establish (iv) and (v). If $f = G^{\sim}_{1}g$ with g in $L^{2}(\underline{\underline{X}}, dx)$ let $\gamma f = R^{\sim}_{(1)}\,\pi^{\sim}_{1}\,g$. Clearly γf is in $\underline{\underline{H}}^{\sim}$, the difference $f - H^{\sim}_{1}\gamma f$ is in $\underline{\underline{F}}$ and (20.29) is valid. Moreover since

$$\int dx\, H^{\sim}_{1}\gamma f(x)\; GH^{\sim}_{1}\gamma f(x) \;\leq\; \int dx\, H^{\sim}_{1}\gamma f(x)\; H^{\sim}\gamma f(x)$$

$$= U^{\sim}_{0,1}(\gamma f, \gamma f) \leq \int \nu\,(dy)\,\{\gamma f(y)\}^{2}$$

the difference $f - H^{\sim}\gamma f = G_{1}g - GH^{\sim}_{1}\gamma f$ belongs to $\underline{\underline{F}}_{(e)}$ and

$$E(f-H^{\sim}\gamma f, f-H^{\sim}\gamma f) + Q^{\sim}(\gamma f, \gamma f)$$

$$= E_{1}(G_{1}g, f-H^{\sim}\gamma f) - \int dx\; G\;\; g(x)\,\{\, f(x)-H^{\sim}\gamma f(x)\,\}$$

$$-E(GH^{\sim}_{1}\gamma f, f-H^{\sim}\gamma f) + Q^{\sim}\;(\gamma f, \gamma f)$$

$$= \int dx\ \{ g(x) - G_1 g(x) \}\ \{ f(x) - H^{\sim}\gamma f(x) \} - \int dx\ H_1^{\sim}\gamma f(x) \{ f(x) - H^{\sim}\gamma f(x) \}$$

$$+ Q^{\sim}(\gamma f, \gamma f)$$

$$= \int dx\, g(x) \{ f(x) - H^{\sim}\gamma f(x) \} + \int dx\ G_1 g(x)\ H^{\sim}\gamma f(x) - \int dx\ f^2(x)$$

$$+ Q^{\sim}_{(1)}\ (\gamma f, \gamma f)$$

$$= \int dx\, g(x) \{ f(x) - H_1^{\sim}\gamma f(x) \} + Q^{\sim}_{(1)}(\gamma f, \gamma f) - \int dx\ f^2(x)$$

$$= E_1(f - H_1^{\sim}\gamma f, f - H_1^{\sim}\gamma f) + Q^{\sim}_{(1)}(\gamma f, \gamma f) - \int dx\ f^2(x)$$

$$= E_1^{\sim}(f,f) - \int dx\, f^2(x)$$

$$= E^{\sim}\ (f,f)$$

which establishes (20.28) and the first half of (iv) follows after a passage to the limit in f. The first half of (v) is immediate and uniqueness of γ follows directly from (i). The second halves of (iv) and (v) will follow once we show that functions $R^{\sim}_{(1)}\ \pi_1^{\sim} g$ with g in $L^2(dx)$ are dense in $\underline{H}^{\sim}$. But if this is false then there exists nontrivial φ in $\underline{\underline{H}}^{\sim}$ such that

$$Q^{\sim}_{(1)}\ (\varphi\ , R^{\sim}_{(1)}\ \pi_1^{\sim} g) = \int \nu\ (dy) \varphi\ (y)\ \pi_1^{\sim} g(y)$$

$$= \int dx\, g(x)\ H_1^{\sim}\ \varphi\ (x)$$

$$= 0$$

for all g in $L^2(dx)$ and therefore $U^{\sim}_{1,u}\ (\varphi, \varphi) = 0$. But then $U^{\sim}_{1,u}(1, \varphi^2) = U^{\sim}_{1,u} \langle\varphi, \varphi\rangle$ is bounded independent of u which contradicts (i). This completes the proof of the theorem. ///

The next theorem gives some a priori control over possible choices for Δ in Theorem 20.2

Theorem 20.3. Let $(\underline{\underline{F}}^{\sim}, E^{\sim})$ be an extension of $(\underline{\underline{F}}, E)$. After possibly replacing $(\underline{\underline{F}}, E)$ by an equivalent representation we can choose a regularizing space $\underline{\underline{X}}^{\sim}$ for $(\underline{\underline{F}}^{\sim}, E^{\sim})$ and a regularizing space $\underline{\underline{X}}^{env}$ for $(\underline{\underline{F}}_a^{env}, E^{res})$ such that $\underline{\underline{X}}$ is open in both $\underline{\underline{X}}^{\sim}$ and $\underline{\underline{X}}^{env}$ and there exists a continuous surjection $p : \underline{\underline{X}}^{env} \to \underline{\underline{X}}^{\sim}$ which fixes the points of $\underline{\underline{X}}$. ///

Proof. It follows from (20.1) and (20.3) that $\underline{\underline{F}}^{\sim}$ is contained in $\underline{\underline{F}}_a^{env}$. Choose $\underline{\underline{B}}_0^{env}$ satisfying 2.2.1 through 2.2.4 for $\underline{\underline{F}}_a^{env}$ in such a way that $\underline{\underline{B}}_0^{\sim} = \underline{\underline{B}}_0^{env} \cap \underline{\underline{F}}^{\sim}$ and $\underline{\underline{B}}_0 = \underline{\underline{B}}_0^{env} \cap \underline{\underline{F}}$ satisfy the corresponding conditions for $\underline{\underline{F}}^{\sim}$ and $\underline{\underline{F}}$. Then let $\underline{\underline{X}}^{env}$ and $\underline{\underline{X}}^{\sim}$ be the maximal ideal spaces for the uniform closures of $\underline{\underline{B}}_0^{env}$ and $\underline{\underline{B}}_0^{\sim}$ and replace $\underline{\underline{X}}$ by the maximal ideal space for the uniform closure of $\underline{\underline{B}}_0$. ///

Remark. Theorem 20.3 has limited application since in general $\underline{\underline{X}}^{env}$ depends on $(\underline{\underline{F}}^{\sim}, E^{\sim})$ in a nontrivial manner. This is because the preimage of a nonpolar set in $\underline{\underline{X}}^{\sim}$ may be polar in $\underline{\underline{X}}^{env}$ and quasi-homeomorphisms of $\underline{\underline{X}}^{env}$ ignore polar sets. For example if $\underline{\underline{F}}_a^{res} = \underline{\underline{F}}_a^{env}$ then $\underline{\underline{X}}$ itself is a regularizing space for $\underline{\underline{F}}_a^{env}$. Since any proper extension of $(\underline{\underline{F}}, E)$ has a regularizing space which properly contains $\underline{\underline{X}}$ we might be tempted to conclude from Theorem 20.3 that $(\underline{\underline{F}}, E)$ has no proper extensions. This would be a very powerful result but unfortunately it is false as the following example demonstrates. Let dx be ordinary Lebesgue measure on the real line $\underline{\underline{R}}$, let $0 < \alpha < 1$ and let $\varkappa(x)$ be integrable on the complement of any neighborhood of 0 but such that

$$(20.48) \qquad \int_{-1}^{0} dx\, \varkappa(x) = +\infty \; ; \quad \int_{0}^{1} dx\, \varkappa(x) = +\infty \, .$$

Let

$$E(f,f) = \tfrac{1}{2} \int_{-\infty}^{+\infty} dx \int_{-\infty}^{+\infty} dy |x-y|^{-1-\alpha} \{f(x)-f(y)\}^2$$

$$+ \int_{-\infty}^{+\infty} dx\, \kappa(x) f^2(x)$$

and let $\underline{\underline{F}}$ be the collection of f in $L^2(\underline{\underline{R}}, dx)$ such that $E(f,f) < +\infty$. Evidently $(\underline{\underline{F}}, E)$ is a Dirichlet space on $L^2(\underline{\underline{R}}, dx)$. Let $(\underline{\underline{F}}^{env}_a, E^{res})$ be the Dirichlet space associated with a symmetric stable process of $\underline{\underline{R}}$ with index α. That is,

$$E^{res}(f,f) = \tfrac{1}{2} \int_{-\infty}^{+\infty} dx \int_{-\infty}^{+\infty} dy |x-y|^{-1-\alpha} \{f(x)-f(y)\}^2 ,$$

$\underline{\underline{F}}^{env}$ is the collection of f for which $E^{res}(f,f) < +\infty$ and $\underline{\underline{F}}^{env}_a = \underline{\underline{F}}^{env} \cap L^2(dx)$. It is well known and easily checked directly using convolutions that $(\underline{\underline{F}}^{env}_a, E^{res})$ is regular on $\underline{\underline{R}}$. Also a simple computation shows that the indicator of the half line $[0,\infty)$ belongs to $\underline{\underline{F}}^{env}$ and so again convolutions can be used to show that $(\underline{\underline{F}}, E)$ is regular on $\underline{\underline{X}} = \underline{\underline{R}} - \{0\}$. Once this is established it follows directly that $(\underline{\underline{F}}^{env}, E^{res})$ is indeed the enveloping space for $(\underline{\underline{F}}, E)$. By the general criterion of [53, p.7] individual points are polar for $\underline{\underline{F}}^{env}$ and so by Theorems 19.1 and 19.2 $\underline{\underline{F}}^{env} = \underline{\underline{F}}^{res}$; that is, $\underline{\underline{F}}$ is E^{res} dense in $\underline{\underline{F}}^{env}$. This can also be shown directly without recourse to Section 19. We consider two Dirichlet space which are extensions of $(\underline{\underline{F}}, E)$. Let $\underline{\underline{F}}^1$ be the set of f in $\underline{\underline{F}}^{env}_a$ for which there exists one constant $f(0)$ such that

$$\int_{-\infty}^{+\infty} dx\, \kappa(x) \{f(x) - f(0)\}^2 < +\infty$$

and let $\underline{\underline{F}}^2$ be the set of f in $\underline{\underline{F}}^{env}_a$ for which there exist two constants $f(0^-)$, $f(0^+)$ such that

$$\int_{-\infty}^{0} dx\, \varkappa(x) \{ f(x)-f(0^-)\}^2 + \int_{0}^{\infty} dx\, \varkappa(x) \{f(x)-f(0^+)\}^2 < +\infty \tag{20.20}$$

and define

$$E^1(f,f) = E^{res}(f,f) + \int_{-\infty}^{+\infty} dx\, \varkappa(x) \{ f(x)-f(0)\}^2$$

$$E^2(f,f) = E^{res}(f,f) + \int_{-\infty}^{0} dx\, \varkappa(x) \{f(x)-f(0^-)\}^2$$

$$+ \int_{0}^{+\infty} dx\, \varkappa(x) \{ f(x)-f(0^+)\}^2$$

for f in $\underline{\underline{F}}^1, \underline{\underline{F}}^2$ respectively. The constants $f(0)$, $f(0^+)$, $f(0^-)$ are unique when they exist because of (20.48) and it is clear that $(\underline{\underline{F}}^1, E^1)$, $(\underline{\underline{F}}^2, E^2)$ are indeed extensions of $(\underline{\underline{F}}, E)$. Moreover since the indicator of $[0,\infty)$ is in $\underline{\underline{F}}^{env}$, it is clear that $\underline{\underline{F}}^2$ properly contains $\underline{\underline{F}}^1$. In treating $\underline{\underline{F}}^1$ a possible choice of a regularizing space for $\underline{\underline{F}}_a^{env}$ in Theorem 20.3 is $\underline{\underline{X}}^{env,1} = \underline{\underline{R}}$. Then also $\underline{\underline{X}}^1 = \underline{\underline{R}}$ and p is the identity. The origin is polar for $\underline{\underline{F}}^{env}$ but not for $\underline{\underline{F}}^1$. This choice does not work for $\underline{\underline{F}}^2$. Instead one can take $\underline{\underline{X}}^{env,2} = (-\infty, 0^-] \cup [0^+, \infty)$ with the obvious conventions. Now $\underline{\underline{X}}^2 = \underline{\underline{X}}^{env,2}$ and p is the identity and $\{0^-, 0^+\}$ is polar for $\underline{\underline{F}}_a^{env}$ but not $\underline{\underline{F}}^2$. ///

20.4 Terminology Any space Δ together with subprobabilities $\mu(x, d\,)$ as described in the converse part of Theorem 20.2 will be denoted by (Δ, μ) and referred to as a pair. The relevant machinery introduced at the beginning of this section, including Notation 20.3, will always be taken for granted. ///

20.5. Terminology. When $(\underline{\underline{F}}^{\sim}, E^{\sim})$ corresponds to the resolvent constructed in Theorem 20.2 with $(\underline{\underline{H}}^{\sim}, Q^{\sim}) = (\underline{\underline{N}}^{\sim}, N^{\sim})$ we call $(\underline{\underline{F}}^{\sim}, E^{\sim})$ the reflected space modified by (Δ, μ). When $(\underline{\underline{F}}^{\sim}, E^{\sim})$ corresponds to general $(\underline{\underline{H}}^{\sim}, Q^{\sim})$ we say that $(\underline{\underline{F}}^{\sim}, E^{\sim})$ is subordinate to the reflected space modified by (Δ, μ). A given

$(\underline{\underline{F}}^{\sim}, E^{\sim})$ may be subordinate to more than one modified reflected space. We will use this terminology only when $(\underline{\underline{F}}^{\sim}, E^{\sim})$ is an extension of $(\underline{\underline{F}}, E)$. ///

The next theorem is an easy corollary of Theorem 20.2.

Theorem 20.4. *Let $(\underline{\underline{F}}^{\sim}, E^{\sim})$ be an extension of $(\underline{\underline{F}}, E)$ and let $(\underline{\underline{F}}^{ref\sim}, E^{ref\sim})$ be a modified reflected space of $(\underline{\underline{F}}, E)$ which is also an extension of $(\underline{\underline{F}}, E)$. Then $(\underline{\underline{F}}^{\sim}, E^{\sim})$ is subordinate to $(\underline{\underline{F}}^{ref\sim}, E^{ref\sim})$ if and only if $\underline{\underline{F}}^{\sim}$ is contained in $\underline{\underline{F}}^{ref\sim}$ and the difference $E^{\sim} - E^{ref\sim}$ is contractive on $\underline{\underline{F}}^{\sim}$.* ///

The active reflected space $(\underline{\underline{F}}_a^{ref}, E)$ is a special case of a modified reflected space. For the pair we can take $(\Delta^{ref}, 0)$ where

(20.49) $$\Delta^{ref} = \underline{\underline{X}}^{ref} - \underline{\underline{X}}$$

with $\underline{\underline{X}}^{ref}$ an appropriate regularizing space for $(\underline{\underline{F}}_a^{ref}, E)$. This follows easily from our results in Section 14. Now Theorem 20.4 together with Theorem 15.2 (First Structure Theorem) and its obvious converse yield

Theorem 20.5 *A Dirichlet space $(\underline{\underline{F}}^{\sim}, E^{\sim})$ has generator $A^{\sim}$ contained in the local generator $\mathcal{Q}$ if and only if it is subordinate to the active reflected space $(\underline{\underline{F}}_a^{ref}, E)$.* ///

Remark. The proof of Theorem 20.2 can be used to give a proof of Theorem 15.2 which is independent of Fukushima's argument used in Section 15. If $(\underline{\underline{F}}^{\sim}, E^{\sim})$ has generator $A^{\sim}$ contained in the local generator $\mathcal{Q}$ then the arguments at the beginning of Section 15 show that $(\underline{\underline{F}}^{\sim}, E^{\sim})$ is an extension of $(\underline{\underline{F}}, E)$ and the proof of Theorem 20.2 goes through with $\underline{\underline{X}}^{\sim} = \underline{\underline{X}}^{ref}$ and $\mu = 0$. Thus Theorem 15.2 is really a special case of Theorem 20.2-(iii). ///

20.22

We look more carefully now at the structure of modified reflected spaces. First some preliminaries.

Lemma 20.6. If $f \in \underline{F}^{env}$, then for quasi-every x in $\underline{X}$

$$P_x[X_{\zeta-0} = \partial \,;\ \mathrm{Lim}_{t\uparrow\zeta} f(X_t) \text{ doesn't exist}] = 0. \quad ///$$

Proof. This follows from Theorem 18.2-(i) and Theorem 18.1-(iv) and from Theorem 14.4 with $(\underline{\underline{F}}_a^{res}, E^{res})$ playing the role of $(\underline{F}, E)$. ///

Lemma 20.7. Let (Δ, μ) be a pair.

(i) Let D be open with compact closure, let $M = \underline{X} - D$ and let bounded f in $\underline{F}_{loc}$ have a representation

$$f = f^D + N^D \varkappa\mu\,(\cdot, \varphi) + H^M f \tag{20.50}$$

with f^D in the absorbed space $\underline{F}^D$ and with φ bounded on Δ. Then

$$\begin{aligned}(20.51)\ & \tfrac{1}{2}\int_D \langle A_c f\rangle(dx) + \tfrac{1}{2}\iint_{D\times D} J(dx,dw)\{f(x)-f(w)\}^2 \\ & + \int_{D\times M} J(dx,dy)\{f(x)-f(y)\}^2 + \int_D \varkappa(dx)r(x)f^2(x) \\ & + \int_D \varkappa(dx)\int_\Delta \mu(x,dy)\{f(x)-\varphi(y)\}^2 \\ = {} & E(f^D, f^D) + \int_D \varkappa(dx)\int_\Delta \mu(x,dy)\int_M H^M(x,dz)\{\varphi(y)-f(z)\}^2 \\ & + \int_D \varkappa(dx)r(x)\, H^M f^2(x) + \int_D \varkappa(dx)r(x)\int_\Delta N^D\varkappa\mu(x,dy)\varphi^2(y) \\ & + \tfrac{1}{2}\iint_{\Delta\times\Delta} (\mu\varkappa N^D\varkappa\mu)(dy,dz)\{\varphi(y)-\varphi(z)\}^2 \\ & + \tfrac{1}{2}\delta\, \Sigma_i\, I(r(i)<\zeta)\{f(X_{r(i)}) - f(X_{e(i)-0})\}^2\end{aligned}$$

(ii) Let bounded f in $\underline{\underline{F}}_{loc}$ have a representation

$$f = f_o + N\varkappa\mu(\cdot,\varphi) + h \tag{20.52}$$

where φ is bounded on Δ, where $f_o \varepsilon \underline{\underline{F}}_{(e)b}$ and where h is bounded and harmonic and therefore has a representation

$$h(x) = \mathscr{E}_x \Phi$$

with Φ a terminal variable as in Section 14. Then

$$E^{res}(f,f) + \int_{\underline{\underline{X}}} \varkappa(dx) r(x) f^2(x) + \int_{\underline{\underline{X}}} \varkappa(dx) \int_{\Delta} \mu(x,dy)\{f(x)-\varphi(y)\}^2 \tag{20.53}$$

$$= E(f_o,f_o) + \int_{\underline{\underline{X}}} \varkappa(dx) \int_{\Delta} \mu(x,dy) \int_{[X_{\zeta-0}=\partial]} P_x(d\omega)\{\varphi(y)-\Phi(\omega)\}^2$$

$$+ \int_{\underline{\underline{X}}} \varkappa(dx) r(x) \mathscr{E}_x \Phi^2 + \int_{\underline{\underline{X}}} \varkappa(dx) r(x) \int_{\Delta} N\varkappa\mu(x,dy)\varphi^2(y)$$

$$+ \tfrac{1}{2} \iint_{\Delta\times\Delta} (\mu\varkappa N\varkappa\mu)(dy,dz)\{\varphi(y)-\varphi(z)\}^2$$

$$+ \tfrac{1}{2} \mathscr{E}\, I(X_{\zeta^*} = X_{\zeta-0} = \partial)\{\Phi - \Phi\cdot\rho\}^2 . \quad ///$$

In (20.51) and (20.53) we are using the notation of Section 13 and also introducing some new notation which is intended to be more or less transparent. For example $N\varkappa\mu(x,dy)$ is the measure on Δ determined for quasi-every x in $\underline{\underline{X}}$ by

$$\int N\varkappa\mu(x,dy)\varphi(y) = N\{\int \mu(\cdot,dy)\varphi(y)\cdot\varkappa\}(x),$$

$N\varkappa\mu(\cdot,\varphi)$ is the function

$$N\varkappa\mu(x,\varphi) = \int N\varkappa\mu(x,dy)\varphi(y)$$

20.24

and $(\mu\varkappa\ N\varkappa\mu)\ (dy,dz)$ is the measure on $\Delta \times \Delta$ determined by

$$\iint (\mu\varkappa\ N\varkappa\mu)(dy,dz)\ \varphi(y)\psi(z)$$

$$= \int \varkappa(dx)\mu(x,\varphi)\ N\varkappa\mu(x,\psi)$$

$$= \int \varkappa(dx)\mu(x,\psi)\ N\varkappa\mu(x,\varphi).$$

For the proof adjoin variables Y^*,Y to the extended space Ω_∞ such that Y^*,Y takes values in $\Delta \cup \{\partial\}$ and such that conditioned on all the trajectory variables Y^*,Y are mutually independent and depend only on $X_{\zeta^*}, X_{\zeta-0}$ respectively and are determined by

$$Y = \partial \qquad \text{on } [X_{\zeta-0} = \partial]$$

$$P[Y \in \Gamma \mid X_{\zeta-0}] = \mu(X_{\zeta-0},\Gamma) \qquad \text{on } [X_{\zeta-0} \in \underline{\underline{X}}]$$

$$P[Y = \partial \mid X_{\zeta-0}] = r(X_{\zeta-0}) \qquad \text{on } [X_{\zeta-0} \in \underline{\underline{X}}]$$

with an analagous prescription for Y^* depending on X_{ζ^*}. Extend the definition of the time reversal operator ρ by $Y \cdot \rho = Y^*$. Also adjoin Y^*,Y to the standard sample space Ω in an analagous manner except that X_0 plays the role of X_{ζ^*}. For (i) we note that it suffices to consider the special case $f^D = G^D g$ with g bounded and therfore integrable on D, we introduce the notations

$$f^{\sim}(X_{\sigma(M)}) = I(\sigma(M) < +\infty)\ f(X_{\sigma(M)}) + I(\sigma(M) = +\infty)\ \varphi(Y)$$

$$f^{\sim}(X_{r(i)}) = I(r(i) < \zeta)\ f(X_{r(i)}) + I(r(i) = \zeta)\ \varphi(Y)$$

and then we let

$$M^{\sim}f(t) = M^D f(t) + H^M f(X_t) + N^D \varkappa\mu(X_t,\varphi) \text{ for } X_t \in D$$

$$M^{\sim}f(\sigma(M)) = Mf^D(\sigma(M)) + f^{\sim}(X_{\sigma(M)})$$

$$M^{\sim}f(r(i)) = Mf^D(r(i)) + f^{\sim}(X_{r(i)}).$$

20.25

Both on Ω and Ω_∞ these fundtionals are martingales "along excursions into D" and the random measures $\langle M^{\sim}f\rangle(dt)$ and $\langle \tilde{M}_c f\rangle(dt)$ are well defined for the set of t such that $X_t \in D$. Now

(20.54) $$\tfrac{1}{2}\mathcal{E}_x \int_0^{\sigma(M)} \langle M^{\sim}f\rangle(dt)$$

$$= \tfrac{1}{2}\mathcal{E}_x\{f^{\sim}(X_{\sigma(M)}) - f(X_0) + \int_0^{\sigma(M)} dt\, g(X_t)\}^2$$

$$= \mathcal{E}_x \int_0^{\sigma(M)} dt\, g(X_t) G^D g(X_t)$$

$$+ \tfrac{1}{2}\mathcal{E}_x\{f^{\sim}(X_{\sigma(M)}) - \varphi(Y^*)\}^2$$

$$+ \mathcal{E}_x\{f^{\sim}(X_{\sigma(M)}) - \varphi(Y^*)\} \int_0^{\sigma(M)} dt\, g(X_t)$$

$$- \tfrac{1}{2}\mathcal{E}_x\{\varphi(Y^*) - f(X_0)\}^2$$

since

$$\mathcal{E}_x\{\varphi(Y^*) - f(X_0)\}\{Mf(\sigma(M)) - Mf(0)\} = 0.$$

But

$$\mathcal{E}_x\{\varphi(Y^*) - f(X_0)\}^2 = r(x)f^2(x) + \int \mu(x,dy)\{\varphi(y) - f(x)\}^2$$

and so (20.54) can be rewritten

(20.55) $$\tfrac{1}{2}\mathcal{E}_x \int_0^{\sigma(M)} \langle \tilde{M}f\rangle(dt) + \tfrac{1}{2}r(x)f^2(x) + \tfrac{1}{2}\int \mu(x,dy)\{f(x) - \varphi(y)\}^2$$

$$= \mathcal{E}_x \int_0^{\sigma(M)} dt\, g(X_t)\, G^D g(X_t)$$

$$+ \tfrac{1}{2}\mathcal{E}_x\{f^2(X_{\sigma(M)}) - \varphi(Y^*)\}^2$$

$$+ \mathcal{E}_x\{f^2(X_{\sigma(M)}) - \varphi(Y^*)\} \int_0^{\sigma(M)} dt\, g(X_t).$$

Let $e(i), \rho(i)$ be approximating times and let $e(i)$ be temporarilly relabeled as in Section 13. Then

20.26

$$\tfrac{1}{2}\mathcal{E}_x \int_{\varepsilon(i)}^{\rho(i)} \langle M^{\sim} f\rangle (dt) + \tfrac{1}{2}\mathcal{E}_x \{f(X_{\varepsilon(i)}) - f(X_{\varepsilon(i)-0})\}^2$$

$$= \tfrac{1}{2}\mathcal{E}_x \{ f^{\sim}(X_{\rho(i)}) - f(X_{\varepsilon(i)}) + \int_{\varepsilon(i)}^{\rho(i)} dt\, g(X_t)\}^2$$

$$= \mathcal{E}_x \int_{\varepsilon(i)}^{\rho(i)} dt\, g(X_t)\, G^D g(X_t)$$

$$+ \tfrac{1}{2}\mathcal{E}_x \{f^{\sim}(X_{\rho(i)}) - f(X_{e(i)-0})\}^2$$

$$+ \mathcal{E}_x \{f^{\sim}(X_{\rho(i)}) - f(X_{e(i)-0})\} \int_{\varepsilon(i)}^{\rho(i)} dt\, g(X_t).$$

Summing over i and passing to the limit $D' \uparrow D$ we get

$$(20.56) \quad \tfrac{1}{2}\mathcal{E}_x \int_{\sigma(M)}^{\zeta} \langle M^{\sim} f\rangle (dt)\, 1_D(X_t) + \tfrac{1}{2}\mathcal{E}_x \Sigma_i \{f(X_{e(i)}) - f(X_{e(i)-0})\}^2$$

$$= \mathcal{E}_x \int_{\sigma(M)}^{\zeta} dt\, g(X_t) G^D g(X_t)$$

$$+ \tfrac{1}{2}\mathcal{E}_x \Sigma_i \{f^{\sim}(X_{r(i)}) - f(X_{e(i)-0})\}^2$$

$$+ \mathcal{E}_x \Sigma_i \{f^{\sim}(X_{r(i)}) - f(X_{e(i)-0})\} \int_{e(i)}^{r(i)} dt\, g(X_t)$$

Combining (20.55) with (20.56), passing to the process $\mathcal{P}$ and eliminating the cross term with the help of time reversal as in Section 13, we get

$$(20.57) \quad \tfrac{1}{2}\mathcal{E} \int_{\zeta^*}^{\zeta} \langle M^{\sim} f\rangle (dt)\, 1_D(X_t) + \tfrac{1}{2}\int_D \varkappa(dx) r(x) f^2(x)$$

$$+ \tfrac{1}{2}\int_D \varkappa(dx) \int \mu(x, dy) \{f(x) - \varphi(y)\}^2 + \tfrac{1}{2}\mathcal{E}\Sigma_i \{f(X_{e(i)}) - f(X_{e(i)-0})\}^2$$

$$= E(f^D, f^D) + \tfrac{1}{2}\mathcal{E} I(X_{\zeta^*} \in D)\{f^{\sim}(X_{\sigma(M)}) - \varphi(Y^*)\}^2$$

$$+ \tfrac{1}{2}\mathcal{E}\Sigma_i \{f^{\sim}(X_{r(i)}) - f(X_{e(i)-0})\}^2.$$

With the help of Meyer's theory on the decomposition of square integrable martingales [35, Chap. VIII] it is easy to check that

$$\tfrac{1}{2}\mathcal{E}\int_{\zeta^*}^{\zeta} \langle M^{\sim}f\rangle\,(dt)\,1_D(X_t)$$

$$= \tfrac{1}{2}\int_D \langle A_c f\rangle\,(dx) + \tfrac{1}{2}\iint_{D\times D} J(dx,dw)\,\{f(x)-f(w)\}^2$$

$$+ \tfrac{1}{2}\iint_{D\times M} J(dx,dy)\,\{f(x)-f(y)\}^2$$

$$+ \tfrac{1}{2}\int_D \kappa(dx)r(x)f^2(x) + \tfrac{1}{2}\int_D \kappa(dx)\int \mu(x,dy)\{f(x)-\varphi(y)\}^2 .$$

Also it is clear that

$$\tfrac{1}{2}\mathcal{E}\,\Sigma_i\{f(X_{e(i)})-f(X_{e(i)-0})\}^2$$

$$=\tfrac{1}{2}\iint_{D\times M} J(dx,dy)\{f(x)-f(y)\}^2$$

$$\tfrac{1}{2}\mathcal{E}\,I(X_{\zeta^*}\varepsilon\, D)\;\{f^{\sim}(X_{\sigma(M)})-\varphi(Y^*)\}^2$$

$$=\tfrac{1}{2}\int_D \varkappa(dx)r(x)\,H^M f^2(x)+\tfrac{1}{2}\int_D \varkappa(dx)r(x)\int N^D\varkappa\mu(x,dy)\varphi^2(y)$$

$$+\tfrac{1}{2}\int_D \varkappa(dx)\int\mu(x,dy)\int H^M(x,dz)\{\varphi(y)-f(z)\}^2$$

$$+\tfrac{1}{2}\iint(\mu\varkappa N^D\varkappa\mu)(dy,dz)\{\varphi(y)-\varphi(z)\}^2$$

$$+\tfrac{1}{2}\int_D \varkappa(dx)\int\mu(x,dy)\varphi^2(y)\;N^D r\cdot\varkappa(x)$$

$$\tfrac{1}{2}\mathcal{E}\,\Sigma_i\{f^{\sim}(X_{r(i)})-f(X_{e(i)-0})\}^2$$

$$=\tfrac{1}{2}\mathcal{E}\,\Sigma_i I(r(i)<\zeta)\;\{f(X_{r(i)})-f(X_{e(i)-0})\}^2$$

$$+\tfrac{1}{2}\int_D \varkappa(dx)r(x)\,H^M f^2(x)$$

$$+\tfrac{1}{2}\int_D \varkappa(dx)\int\mu(x,dy)\int H^M(x,dz)\{\varphi(y)-f(z)\}^2$$

and Lemma 20.7-(i) follows. For (ii) assume $f=Gg$ with g bounded and integrable and define

$$M^{\sim}f(t) = Mf_o(t) + h(X_t) + N\varkappa\mu(X_t,\omega) \qquad \text{for } X_t \in \underline{\underline{X}}$$

$$M^{\sim}f(t) = Mf_o(t) + \Phi + \varphi(Y) \qquad \text{for } t \geq \zeta$$

$$M^{\sim}f(t) = Mf_o(t) + \Phi\circ\rho + \varphi(Y^*) \qquad \text{for } t < \zeta^*.$$

Then

(20.58) $$\tfrac{1}{2}\mathcal{E}_x < M^{\sim}f > (\zeta)$$
$$= \tfrac{1}{2}\,\mathcal{E}_x\{\Phi + \varphi(Y) + \int_0^{\zeta} dt\, g(X_t)\}^2$$
$$= \mathcal{E}_x \int_0^{\zeta} dt\, g(X_t)\, Gg(X_t) + \tfrac{1}{2}\,\mathcal{E}_x\{\Phi + \varphi(Y) - f(X_0)\}^2$$
$$+ \mathcal{E}_x\{\Phi + \varphi(Y) - f(X_0)\} \int_0^{\zeta} dt\, g(X_t)$$

Integrating with respect to $L_k^{\circ}(dx)$ and letting $k \uparrow \infty$ as in Section 14 we get

(20.59) $$\tfrac{1}{2}\mathcal{E}\, I(X_{\zeta^*} = \partial) \int_{\zeta^*}^{\zeta} < M^{\sim}f > (dt)$$
$$= \mathcal{E}\, I(X_{\zeta^*} = \partial) \int_{\zeta^*}^{\zeta} dt\, g(X_t) Gg(X_t)$$
$$+ \tfrac{1}{2}\,\mathcal{E}\, I(X_{\zeta^*} = \partial)\ \{\Phi + \varphi(Y) - \Phi\cdot\rho\}^2$$
$$+ \mathcal{E} I(X_{\zeta^*} = \partial)\ \{\Phi + \varphi(Y) - \Phi\cdot\rho\} \int_{\zeta^*}^{\zeta} dt\, g(X_t).$$

Replacing (20.58) by

(20.58') $$\tfrac{1}{2}\mathcal{E}_x < M^{\sim}f > (\zeta) + \tfrac{1}{2} r(x) f^2(x) + \tfrac{1}{2}\int \mu(x,dy)\{\varphi(y) - f(x)\}^2$$
$$= \mathcal{E}_x \int_0^{\zeta} dt\, g(X_t) Gg(X_t) + \tfrac{1}{2}\mathcal{E}_x\{\Phi + \varphi(Y) - \varphi(Y^*)\}^2$$
$$+ \mathcal{E}_x\{\Phi + \varphi(Y) - \varphi(Y^*)\} \int_0^{\zeta} dt\, g(X_t),$$

integrating (20.58') with respect to $\varkappa(dx)$ and combining with (20.59) we get after eliminating the cross term

(20.60) $$\tfrac{1}{2}\mathcal{E}\int_{\zeta^*}^{\zeta} < M^{\sim}f > (dt) + \tfrac{1}{2}\int \varkappa(dx) r(x) f^2(x)$$
$$+ \tfrac{1}{2}\int \varkappa(dx) \int \mu(x,dy)\{f(x) - \varphi(y)\}^2$$
$$= E(f_0, f_0) + \tfrac{1}{2}\mathcal{E}\,\{\Phi + \varphi(Y) - \Phi\cdot\rho - \varphi(Y^*)\}^2.$$

Conclusion (ii) follows since

$$\tfrac{1}{2}\mathcal{E}\int_{\zeta^*}^{\zeta} < M^{\sim} f > (dt) = E^{res}(f,f) + \tfrac{1}{2}\int \varkappa\,(dx) r(x) f^2(x)$$

$$+ \tfrac{1}{2}\int \varkappa\,(dx)\int \mu\,(x,dy)\{\,f(x)-\varphi\,(y)\}^2$$

$$\tfrac{1}{2}\mathcal{E}\{\Phi + \varphi\,(Y) - \Phi\cdot\rho - \varphi\,(Y^*)\}^2$$

$$= \int \varkappa\,(dx) r(x)\,\mathcal{E}_x\,\Phi^2 + \int \varkappa\,(dx) r(x) \int N\varkappa\mu(x,dy)\,\varphi^2(y)$$

$$+ \int \varkappa\,(dx)\int \mu\,(x,dy)\int \theta_x(dw)\{\varphi\,(y) - \Phi\,(w)\}^2$$

$$+ \tfrac{1}{2}\iint (\mu\varkappa\, N\varkappa\mu)(dy,dz)\{\varphi\,(y) - \varphi\,(z)\}^2$$

$$+ \tfrac{1}{2}\mathcal{E}\, I(X_{\zeta^*} = X_{\zeta-0} = \partial)\,\{\Phi - \Phi\cdot\rho\}^2 \quad ///$$

Now we are ready for

Theorem 20.8. Let $(\underline{F}^{\sim}, E^{\sim})$ be the reflected space modified by a pair (Δ, μ) satisfying condition (i) of Theorem 20.2. Then for each f in $\underline{F}^{\sim}$ there is a unique function γf in $L^2(\Delta, \nu)$ such that

$$E^{\sim}(f,f) = E^{res}(f,f) + \int \varkappa\,(dx) r(x) f^2(x) + \int \varkappa\,(dx)\int \mu\,(x,dy)\{f(x) - \gamma f(y)\}^2 \qquad (20.61)$$

and such that for quasi-every x in $\underline{X}$

$$P_x[X_{\zeta-0} = \partial\ ;\ \lim_{t\uparrow\zeta} f(X_t) \neq \gamma f(\tilde{X}_{\zeta-0})\,] = 0. \qquad (20.62)$$

Moreover for each f in $\underline{F}_a^{env}$ there is at most one function γf in $L^2(\Delta, \nu)$ satisfying (20.62) such that the right side of (20.61) converges and $\underline{F}^{\sim}$ is the totality of f in $\underline{F}_a^{env}$ for which such γf in $L^2(\nu)$ exists. ///

20.31

Proof. The uniqueness of γf follows since if φ on Δ satisfies

$$\mathcal{E}_x[X_{\zeta-0} = \partial ; \varphi(X^\sim_{\zeta-0}) \neq 0] = 0$$

$$\int \varkappa(dx) \int \mu(x, dy)\, \varphi^2(y) < +\infty$$

then

$$U^\sim_{0,\infty}(1,\varphi^2)$$

$$= \operatorname{Lim}_{v\uparrow\infty} v \int dx(1-vG_v)\, H^\sim 1\, H^\sim \varphi^2(x)$$

$$\leq \operatorname{Lim}_{v\uparrow\infty} v \int dx(1-vG_v) N\varkappa\mu(x,\varphi^2)$$

$$= \int \varkappa(dx)\mu(x,\varphi^2)$$

which violates condition (i) of Theorem 20.2 unless φ is trivial. If $f \in \underline{F}^\sim$ then by Theorem 20.2 there exists γf in $\underline{N}^\sim \subset L^2(\nu)$ such that

$$f = f_o + H^\sim \gamma f$$

$$= f_o + \mathcal{E}_x I(X_{\zeta-0} = \partial)\gamma f(X^\sim_{\zeta-0}) + N\varkappa\, \mu(\cdot, \gamma f)$$

with $f_o \in \underline{E}_{(e)}$. It suffices to consider the case where f and φ and therefore f_o are bounded. Certainly (20.62) is satisfied. Also the hypotheses of Lemma 20.7 are satisfied with $\Phi = I(X_{\zeta-0} = \partial)\,\gamma f(X^\sim_{\zeta-0})$. By Theorem 20.2

$$E^\sim(f,f) = E(f_o,f_o) + N^\sim(\gamma f, \gamma f)$$

$$= E(f_o,f_o) + \tfrac{1}{2}\iint \Theta^\sim(dy,dz)\,\{\gamma f(y) - \gamma f(z)\}^2$$

$$+ \mathcal{E}\, I(X^\sim_{\zeta*} = \partial^\sim)\,\{\gamma f^\sim(X^\sim_{\zeta-0})\}^2$$

$$+ \int \varkappa(dx) r(x)\, H^\sim(\gamma f)^2(x).$$

The last expression agrees with the right side of (20.53) and so (20.61) follows from Lemma 20.7-(ii). Now consider $f \in \underline{F}^{env}_a$ such that γf exists satisfying

(20.62) and such that the right side of (20.61) converges. By the usual arguments we can restrict attention to f and therefore γf bounded. For D as in Lemma 20.7-(i) f certainly has a representation (20.50) with $\varphi = \gamma f$ and it follows from (20.51) that $E(f^D, f^D)$ is bounded independent of D. Thus we can pass to the limit in D and conclude that f has a representation (20.52), also with $\varphi = \gamma f$. By (20.62)

$$h(x) = \mathcal{E}_x I(X_{\zeta-0} = \partial)\, \gamma f(X^{\sim}_{\zeta-0})$$

and the remainder of theorem follows from Lemma 20.7-(ii). ///

We finish with some miscellaneous remarks that will be useful below.

Remark 1. One immediate consequence of our results is that if $\kappa = 0$ then every Dirichlet space $(\underline{F}^{\sim}, E^{\sim})$ which is an extension of $(\underline{F}, E)$ has generator $A^{\sim}$ contained in $\mathcal{A}$. Our results in Section 21 will show more generally that this is true when the associated semigroup $\tilde{P}_t$ dominates P_t. If κ is bounded then there is no loss of generality in taking $\Delta = \Delta^{ref}$. In particular if κ is bounded and if $\underline{F} = \underline{F}^{ref}_a$, then $(\underline{F}, E)$ has no proper extensions. ///

Remark 2. In the special case when $H^{\sim}\underline{H}^{\sim}$ is contained in $L^2(\underline{X}, dx)$, a function f belongs to the domain of the generator $\tilde{A}$ and $A^{\sim} f = g$ if and only if

$$E^{\sim}(f, h) = -\int dx\, g(x) h(x) \tag{20.63}$$

for h in $\underline{F}$ and also

$$Q^{\sim}(\gamma f, \varphi) = -\int dx\, g(x) H^{\sim}\varphi(x) \tag{20.64}$$

for φ in $\underline{H}^{\sim}$. In the general case we must replace (20.64) by the modified condition

$$Q^{\sim}_{(1)}(\gamma f, \varphi) = \int dx\, \{ f(x) - g(x) \}\, H^{\sim}_1 \varphi(x). \tag{20.64'}$$

We introduce now a special notation which will be useful for some of the examples considered below. The normal derivative $\mathcal{D}_n^{\sim} f$ is the functional defined on φ in $\underline{\underline{N}}^{\sim}$ by

(20.65) $\mathcal{D}_n^{\sim} f(\varphi) = - N^{\sim}(\gamma f, \varphi) - \int dx\, g(x)\, H^{\sim}\varphi(x)$

when it makes sense and otherwise by

(20.67') $\mathcal{D}_n^{\sim} f(\varphi) = - N_{(1)}^{\sim}(\gamma f, \varphi) + \int dx\, \{f(x) - g(x)\}\, H_1^{\sim}\varphi(x)$

which is equivalent when (20.57) makes sense. Then (20.67) or (20.67') are equivalent to

(20.68) $\mathcal{D}_n^{\sim} f(\varphi) = Q^{\sim}(\gamma f, \varphi) - N^{\sim}(\gamma f, \varphi).$

We will write $\mathcal{D}_n f$ in place of $\mathcal{D}_n^{\sim} f$ in the special case when $(\Delta, \mu) = (\Delta^{ref}, 0)$. ///

Remark 3. Every f in $\underline{F}$ is automatically in $L^2(\varkappa)$. However if f satisfies

$$f \geq 0; \quad f \in \text{domain } A; \quad Af \in L^1(dx)$$

then also $f \in L^1(\varkappa)$: To see this it suffices to consider bounded measures $\varkappa_n$ increasing to $\varkappa$ and to note that

$$\int \varkappa_n(dx) f(x) = E(f, N\varkappa_n) = - \int dx\, Af(x)\, N\varkappa_n(x)$$

and therefore

$$\int \varkappa\,(dx) f(x) \leq \int dx\, |Af(x)| \quad ///$$

Remark 4. If $\varphi \geq 0$ in $\underline{\underline{N}}^{\sim}$ is nontrivial and if $P_x[X_{\zeta-0} = \partial;\ \varphi(X_{\zeta-0}^{\sim}) \neq 0] = 0$ for quasi-every x then

(20.69) $H^{\sim}\varphi = N\varkappa\mu(\cdot, \varphi)$

and it is easy to check that condition (i) of Theorem 20.2 requires

(20.70) $\int \varkappa(dx)\mu(x, \varphi)\, N^{\varkappa}\mu(x, \varphi) = +\infty.$ ///

21. Third Structure Theorem

In this section we study the most general symmetric submarkovian semigroup $P_t^{\sim}, t > 0$ on $L^2(\underline{X}, dx)$ which dominates $P_t, t > 0$. That is, for $f \geq 0$

$$(21.1) \qquad \tilde{P}_t f \geq P_t f$$

of course in the almost everywhere sense. Fix one such semigroup and let $(\underline{F}^{\sim}, E^{\sim})$ be the corresponding Dirichlet space. From Lemma 1.1 follows immediately

$$(21.2) \qquad \underline{F} \subset \underline{F}^{\sim}$$

$$(21.3) \qquad E^{\sim}(f,f) \leq E(f,f) \qquad f \geq 0 \text{ in } \underline{F}.$$

Moreover the estimate (20.43) shows that $\underline{F}_b$ is an ideal in $\tilde{\underline{F}}_b$ and therefore $\underline{F}^{\sim}$ has a regularizing space $\tilde{\underline{X}}$ such that $\underline{X}$ is densely imbedded as an open subset. (Again it may first be necessary to replace $(\underline{F}, E)$ by an equivalent Dirichlet space.) We take this represetnation for granted now and use the notation of earlier sections. For nonegative φ, ψ in $\underline{F} \cap C_{com}(\underline{X})$ having disjoint supports

$$E(\varphi, \psi) = - \iint J(dx, dw)\, \varphi(x)\, \psi(w)$$

and therefore

$$(21.4) \qquad \iint J(dx, dw)\varphi(x)\psi(w) = \operatorname{Lim}_{t \downarrow 0} (1/t) \int dx\, P_t\varphi(x)\, \psi(x)$$

which together with a corresponding relation for $\tilde{J}$ and $\tilde{P}_t$ guarantees

$$(21.5) \qquad J^{\sim}(dx, dw) \geq J(dx, dw) \qquad \text{on } \underline{X} \times \underline{X}.$$

From Lemma 11.1 it follows directly that

$$(21.6) \qquad \kappa(dx) \geq \kappa^{\sim}(dx) \qquad \text{on } \underline{X}.$$

Also with the help of (11.23) and the argument which follows it is easy to check that

$$(21.7) \qquad \langle \tilde{A}_c f \rangle (dx) \geq \langle A_c f \rangle (dx) \qquad \text{on } \underline{X}$$

for f in $\underline{\underline{F}}$. From (21.3) it follows that

(21.8) $$\int \varkappa^{\sim}(dy)f^2(y) + \tfrac{1}{2}\iint J^{\sim}(dy,dz)\{f(y)-f(z)\}^2 + \int <A_C^{\sim}f>(dy)$$
$$\leq \int \varkappa(dx)f^2(x) + \tfrac{1}{2}\iint J(dx,dw)\{f(x)-f(w)\}^2 + \int <A_Cf>(dx)$$

and so

(21.9) $$\int_{\underline{\underline{X}}} <A_C^{\sim}f>(dx) - \int_{\underline{\underline{X}}} <A_Cf>(dx) \leq \int_{\underline{\underline{X}}} \varkappa(dx)f^2(x)$$

for $f \geq 0$ in $\underline{\underline{F}}$ and also

(21.10) $$J^{\sim}(A,\underline{\underline{X}}^{\sim}) \leq J(A,\underline{\underline{X}}) + \varkappa(A)$$

for A a Borel subset of $\underline{\underline{X}}$. Our first result is

Lemma 21.1. $<A_Cf>(dx) = <A_C^{\sim}f>(dx)$ for f in $\underline{\underline{F}}$. ///

Proof. It follows from (21.9) that there exists a bounded symmetric linear operator D on $L^2(\underline{\underline{X}},\varkappa)$ such that

$$\int \varkappa(dx)Df(x)f(x) = \int_{\underline{\underline{X}}} <A_C^{\sim}f>(dx) - \int_{\underline{\underline{X}}} <A_Cf>(dx)$$

for f in $\underline{\underline{F}}$. The lemma will be proved if we can show that D is a multiplication operator:

(21.11) $$D\varphi(x) = d(x)\varphi(x)$$

with d(x) bounded and Borel measurable. We prove (21.11) using the fact that D is local:

(21.12) $$\int \varkappa(dx)\, D\varphi(x)\psi(x) = 0$$

whenever φ, ψ have disjoint compact closures. Of course this property is an immediate consequence of Theorem 11.9. It follows directly from (21. 12) that

$D1_A\varphi = 1_A D 1_A \varphi$ whenever A is compact and since D is symmetric actually

(21.13) $$D1_A \varphi = 1_A D\varphi$$

for φ in $L^2(\varkappa)$. Indeed an elementary passage to the limit in (21.13) yields

(21.13') $$Dq\varphi = qD\varphi$$

for q bounded and Borel measurable. Fix φ_0 in $L^2(\varkappa)$ such that $\varphi_0 \neq 0$ [a.e.$\varkappa$] and let $d(x) = D\varphi_0(x)/\varphi_0(x)$. Then for general φ in $L^2(\varkappa)$

$$\begin{aligned} D\varphi &= \operatorname{Lim}_{n\uparrow\infty} D\,\tau_n(\varphi/\varphi_0)\,\varphi_0 \\ &= \operatorname{Lim}_{n\uparrow\infty} \tau_n(\varphi/\varphi_0)\, D\varphi_0 \\ &= \operatorname{Lim}_{n\uparrow\infty} \tau_n(\varphi/\varphi_0)\, d(x)\, \varphi_0(x) \\ &= d(x)\varphi(x) \end{aligned}$$

in the $L^2(\varkappa)$ sense and the lemma is proved. ///

From (21.10) follows the existence of subprobabilities $\lambda(x, d\cdot)$ on $\underline{\underline{X}}$ satisfying the usual regularity conditions such that

(21.14) $$\tilde{J}(dx, dw) = J(dx, dw) + \varkappa(dx)\lambda(x, dw) \qquad \text{on } \underline{\underline{X}} \times \underline{\underline{X}}$$

and for f in $\underline{\underline{F}}$

(21.15) $$\tilde{E}(f,f) = E^{res}(f,f) + \tfrac{1}{2}\int_{\underline{\underline{X}}} \varkappa(dx) \int_{\underline{\underline{X}}} \lambda(x, dw)\{f(x)-f(w)\}^2 + \int_{\underline{\underline{X}}} \varkappa(dx)\rho(x)\, f^2(x)$$

where

(21.16) $$\rho(x) \geq 0\ ; \quad \lambda(x) + \rho(x) \leq 1 \qquad [\text{a.e.}\varkappa(dx)].$$

Let $\underline{\underline{F}}^o$ be the $\tilde{E}$ closure of $\underline{\underline{F}}$ in $\tilde{\underline{\underline{F}}}$. The proof of Theorem 18.2 is easily adapted to show that the corresponding semigroup P_t^o, $t > 0$ is the minimal nonnegative solution of

(21.17) $$P_t^o \varphi(x) = P_t\varphi(x) + \mathcal{E}_x I(\zeta \le t;\ X_{\zeta-0} \in \underline{\underline{X}}) \int_{\underline{\underline{X}}} \lambda\,(X_{\zeta-0}, dw)\, P_{t-\zeta}^o\, \varphi(w)\,.$$

Of course $(\underline{\underline{F}}^\sim, E^\sim)$ is an extension of $(\underline{\underline{F}}^o, E^o)$. The converse is clear and we have proved

<u>Theorem 21.2.</u> (<u>Third Structure Theorem</u>) <u>Let</u> $P_t^\sim$, $t > 0$ <u>be a symmetric submarkovian semigroup on</u> $L^2(\underline{\underline{X}}, dx)$ <u>which dominates</u> P_t, $t > 0$ <u>and let</u> $(\underline{\underline{F}}^\sim, E^\sim)$ <u>be the associated Dirichlet space</u>. <u>Then there exists</u> $\rho(x)$, $\lambda(x, d\cdot)$ <u>satisfying</u> (21.16) <u>and also the symmetry condition</u> $\varkappa(dx)\,\lambda(x, dw) = \int \varkappa(dw)\lambda(w, dx)$ <u>on</u> $\underline{\underline{X}} \times \underline{\underline{X}}$ <u>such that if</u> P_t^o <u>is the minimal solution of</u> (21.17) <u>and if</u> $(\underline{\underline{F}}^o, E^o)$ <u>is the associated Dirichlet space on</u> $L^2(\underline{\underline{X}}, dx)$, <u>then</u> $(\underline{\underline{F}}^\sim, E^\sim)$ <u>is an extension of</u> $(\underline{\underline{F}}^o, E^o)$. <u>Conversely if</u> $\rho(x)$, $\lambda(x, d\cdot)$ <u>and</u> $(\underline{\underline{F}}^o, E^o)$ <u>are as specified above and if</u> $(\underline{\underline{F}}^\sim, E^\sim)$ <u>is an extension of</u> $(\underline{\underline{F}}^o, E^o)$ <u>then the associated semigroup</u> $P_t^\sim$, $t > 0$ <u>dominates</u> P_t, $t > 0$. ///

<u>Remark</u>. It is not true in general that if F_b is an ideal in $\underline{\underline{F}}_b^\sim$ and if $E(f,f) \ge E^\sim(f,f)$ for f in $\underline{\underline{F}}$, then the associated semigroup $\underline{\underline{P}}_t^\sim$, $t > 0$ dominates P_t, $t > 0$. This is clear from Lemma 21.1.///

Chapter IV. Examples

The point of Sections 22 through 26 is to illustrate the structure theorems.

In Section 22 the given Dirichlet space corresponds to absorbing barrier Brownian motion on a bounded interval. Since $\kappa = 0$ every extension and indeed every dominating semigroup has generator contained in the local generator. Thus the classification is covered by Feller's results [53]. We rederive the relevant portion using Dirichlet spaces.

A nontrivial killing measure is added in Section 23. Then extensions exist for which the generator is not contained in the local generator and dominating semigroups exist which are not extensions.

The unbounded interval $[0, \infty)$ is the state space in Section 24. If κ is unbounded near ∞ then extensions exist for which the particle can jump back and forth from ∞. When κ is bounded near ∞ no such extensions exist. The appropriate construction in Section 20 yields a dominating semigroup which is not an extension. The net effect is to add an intensity for jumping within the original state space. The point ∞ plays no role for the process. We call it an "illusory point."

In Section 25 the given Dirichlet space is the absorbed process relative to an open interval for a symmetric infinitely divisible process in one dimension. A thorough study must await the resolution of certain technical problems which we have not really addressed ourselves to. But some results can be established by elementary arguments. Nontrivial extensions always exist if there is diffusion present. With no diffusion present, nontivial extensions exist if and only if the Levy measure has infinite first moment in a neighborhood of the origin.

In Section 26 the given Dirichlet space corresponds to the minimal solution of the Kolmogorov equations. Other symmetric solutions are classified by Dirichlet spaces on one given boundary Δ. It is shown in [47] that a possible choice for Δ is the active extremal Martin boundary. To classify stable symmetric chains which do not satisfy the Kolmogorov equations, the boundary Δ must be allowed to vary in general and an intensity $\mu(x,d\cdot)$ must be prescribed for jumping to Δ instead of to the dead point. We have not attempted a systematic investigation of the possibilities for Δ. It seems to us unlikely that the Martin's boundary would be particularly relevant for any such investigation.

Section 27 has little to do with structure theory. In [48] we developed a technique for extending Feller's speed measure-road map decomposition to general symmetric Markov chains, possibly with instantaneous states. The main result is restated here. Then we show that the road map for the relevant enveloping space can be derived directly from the given road by suppressing incomplete excursions from finite sets and then passing to the limit.

22. Diffusions with Bounded Scale; No Killing

Let $\underline{\underline{W}}([0,1])$ be the collection of functions f defined and absolutely continuous on the closed interval $[0,1]$ such that the derivative f' is square integrable and for f in $\underline{\underline{W}}([0,1])$ put

$$W(f,f) = \tfrac{1}{2}\int_0^1 dx\,\{f'(x)\}^2. \tag{22.1}$$

Let $\underline{\underline{W}}_{abs}([0,1])$ be the subcollection of f in $\underline{\underline{W}}([0,1])$ satisfying the boundary conditions

$$f(0) = f(1) = 0. \tag{22.2}$$

In this section $\underline{\underline{X}} = (0,1)$, the reference measure dx is normalized Lebesgue measure on $(0,1)$ and $(\underline{\underline{F}},E) = (\underline{\underline{W}}_{abs}([0,1]), W)$, the Dirichlet space associated with absorbing barrier Brownian motion.

A function f belongs to the domain of the local generator $\mathfrak{a}$ if f is continuously differentiable and if the derivative f' belongs to $\underline{\underline{W}}$ and then

$$\mathfrak{a}f = \tfrac{1}{2} f''. \tag{22.3}$$

Since killing measure $\kappa = 0$ every symmetric semigroup which dominates P_t has generator $\tilde{A}$ contained in $\mathfrak{a}$. We use the results of Section 20 to classify all such generators.

Clearly $\Delta^{ref} = \{0,1\}$ is the only possible choice and $\underline{\underline{N}}$ is the two dimensional vector space of functions φ defined on Δ^{ref} and

$$N(\varphi,\varphi) = \tfrac{1}{2}\{\varphi(0) - \varphi(1)\}^2. \tag{22.4}$$

Let $\{e_0, e_1\}$ be the basis for $\underline{\underline{N}}$ determined by

$$e_0(0) = e_1(1) = 1; \quad e_0(1) = e_1(0) = 0. \tag{22.5}$$

Then

$$He_0(x) = 1-x \quad ; \quad He_1(x) = x. \tag{22.6}$$

There are four nontrivial possibilities for $\underline{\underline{H}}^{\sim}$:

$$(22.7)\qquad \begin{aligned} \underline{\underline{H}}^{\sim} &= \underline{\underline{N}} \\ \underline{\underline{H}}^{\sim} &= \underline{\underline{N}}_0 = \{\varphi \in \underline{\underline{N}} : \varphi(1) = 0\} \\ \underline{\underline{H}}^{\sim} &= \underline{\underline{N}}_1 = \{\varphi \in \underline{\underline{N}} : \varphi(0) = 0\} \\ \underline{\underline{H}}^{\sim} &= \underline{\underline{N}}_p = \{\varphi \in \underline{\underline{N}} : \varphi(0) = \varphi(1)\} \end{aligned}$$

For $\underline{\underline{H}}^{\sim} = \underline{\underline{N}}$ clearly $\underline{\underline{F}}^{\sim} = \underline{\underline{W}}([0,1])$ and the most general possible $Q^{\sim}$ is

$$(22.8)\qquad Q^{\sim}(\varphi,\varphi) = N(\varphi,\varphi) + a\varphi^2(0) + b\varphi^2(1) + c\{\varphi(0) - \varphi(1)\}^2$$

with $a,b,c \geq 0$. It is easy to check that

$$(22.9)\qquad \mathcal{D}_n f(e_0) = \tfrac{1}{2} f'(0);\ \mathcal{D}_n f(e_1) = -\tfrac{1}{2} f'(1).$$

Thus f in the domain of $\mathcal{O}$ belongs to the domain of $A^{\sim}$ if and only if it satisfies the two boundary conditions

$$(22.10)\qquad \tfrac{1}{2} f'(0) = af(0) + c\,\{f(0) - f(1)\}$$

$$(22.10')\qquad -\tfrac{1}{2} f'(1) = bf(1) + c\,\{f(1) - f(0)\}.$$

For $a=b=c=0$ we get the reflected space which corresponds to classical Brownian motion with reflecting barriers. The general case is obtained from this by superimposing killing at the rates a and b at 0 and 1 respectively and jumping between 0 and 1 with intensity c. This interpretation can be made precise either in the context of Chapter II or by a direct construction using Brownian local times in the spirit of [28].

For $\underline{\underline{H}}^{\sim} = \underline{\underline{N}}_0$

$\underline{\underline{F}}^{\sim} = \underline{\underline{W}}_0([0,1]) = \{f \in \underline{\underline{W}}([0,1]) : f(1) = 0\}$

and the most general $Q^{\sim}$ is

$$(22.11)\qquad Q^{\sim}(\varphi,\varphi) = N(\varphi,\varphi) + a\varphi^2(0)$$

with $a \geq 0$. Again $\mathscr{D}_n f(e_0) = \frac{1}{2} f'(0)$ and so the boundary conditions (22.10) are replaced by

(22.12) $$f(1) = 0; \quad \tfrac{1}{2} f'(0) = af(0).$$

For $\underline{\underline{H}}^{\sim} = \underline{\underline{N}}_p$

$$\underline{\underline{F}}^{\sim} = \underline{\underline{W}}_p([0,1]) = \{f \in \underline{\underline{W}}([0,1]) : f(0) = f(1)\}$$

and again the most general $Q^{\sim}$ is (22.11). Now

$$\mathscr{D}_n f(e_0 + e_1) = \tfrac{1}{2}\{ f'(0) - f'(1)\}$$

and the appropriate boundary conditions are

(22.13) $$f(0) = f(1); \quad \tfrac{1}{2}\{ f'(0) - f'(1)\} = af(0).$$

Remark. There seems to be little point in replacing dx by a more general speed measure m. If m is unbounded near either 0 or 1 then every φ in $\underline{\underline{N}}$ vanishes at that point and so it plays no role in our classification scheme. In particular the moment conditions

$$\int_0^{\frac{1}{2}} m(dx)x < +\infty; \quad \int_{\frac{1}{2}}^{1} m(dx)(1-x) < +\infty$$

are irrelevant for us. This is somewhat surprising since if m is unbounded near 0 and 1 but one of these condition is satisfied then P_t is nonconservative and is dominated by submarkovian semigroups other than itself. The point is that none of these are symmetric on $L^2(\underline{\underline{X}}, m)$. ///

23. Diffusions with Bounded Scale; Nontrivial Killing.

$\underline{X}$ and dx are as in Section 22 and $\kappa(dx)$ is a Radon measure on $\underline{X}$. But now

$$\underline{\underline{F}} = \underline{\underline{W}}_{abs}([0,1]) \cap L^2(\kappa)$$

$$E(f,f) = W(f,f) + \int \kappa(dx) f^2(x).$$

Clearly $E^{res} = W$ and $\underline{\underline{F}}_a^{env} = \underline{\underline{W}}([0,1])$. Since $[0,1]$ is the only possible regularizing space for $(\underline{\underline{F}}_a^{env}, E^{res})$ it follows from Theorem 20.3 that there is no possible loss of generality in taking $\Delta = \{0,1\}$ and topologizing $\underline{\underline{X}} \cup \Delta$ as the closed interval $[0,1]$.

A function f belongs to the domain of the local generator $\mathcal{O}$ and $\mathcal{O}f = g$ if the one sided derivatives

$$f^-(x) = \operatorname{Lim}_{h \downarrow 0} \{f(x) - f(x-h)\} / h \tag{23.1}$$

$$f^+(x) = \operatorname{Lim}_{h \downarrow 0} \{f(x+h) - f(x)\} / h$$

exist everywhere on $(0,1]$ and $[0,1)$ respectively and if

$$\tfrac{1}{2} f'(dx) - f(x)\,\kappa(dx) = g(x)dx \qquad \text{on } (0,1). \tag{23.2}$$

The precise meaning of (23.2) is

$$\tfrac{1}{2} f^-(y) - \tfrac{1}{2} f^+(x) - \int_{(x,y)} \kappa(dt) f(t) = \int_x^y dt\, g(t) \tag{23.2'}$$

$$\tfrac{1}{2} f^+(x) - \tfrac{1}{2} f^-(x) - \kappa(\{x\}) f(x) = 0$$

for $x < y$ in $(0,1)$. As a general rule below we simply write (23.2) and understand this to imply the existence of the appropriate derivatives (23.1).

The generators $A^\sim$ contained in $\mathcal{O}$ are classified exactly as in Section 22 except that we must eliminate any boundary near which κ is unbounded. The new feature is that there exist extensions $(\underline{\underline{F}}^{\sim}, E^{\sim})$ with generator $A^{\sim}$ not contained in $\mathcal{O}$ and also there exist Dirichlet spaces $(\underline{\underline{F}}^{\sim}, E^{\sim})$ which are not extensions but the associated semigroup $P_t^{\sim}$ dominates P_t. Before discussing these we take care of some preliminaries.

Let ψ be any solution of the homogeneous equation

(23.3) $$\tfrac{1}{2}\psi'(dx) = \psi(x)\varkappa(dx)$$

which is positive and bounded away from 0 and define

(23.4) $$\eta_0(x) = \psi(x)\int_x^1 dt\, \psi^{-2}(t)$$

$$\eta_1(x) = \psi(x)\int_0^x dt\, \psi^{-2}(t).$$

Both functions satisfy (23.3) and obvious estimates establish

(23.5) $$\eta_0(1) = \eta_1(0) = 0$$

and then by convexity also

(23.6) $$-\infty < \eta_0'(1) \le 0 \quad ; \quad 0 \le \eta_1'(0) < +\infty .$$

The relation

$$\tfrac{1}{2}\eta_1'(t) = \tfrac{1}{2}\eta_1'(\tfrac{1}{2}) + \int_{\frac{1}{2}}^t \varkappa(ds)\,\eta_1(s)$$

establishes

(23.7) $$\eta_0'(1) < -\infty \text{ if and only if } \int_{\frac{1}{2}}^1 \varkappa(ds) < +\infty$$

and a similar relation establishes

(23.7') $$\eta_1'(0) > -\infty \text{ if and only if } \int_0^{\frac{1}{2}} \varkappa(ds) < +\infty .$$

We claim that also

(23.8) $$\eta_1(1) < +\infty \text{ if and only if } \int_{\frac{1}{2}}^{1} \varkappa(ds)\,(1-s) < +\infty$$

$$\eta_0(0) < +\infty \text{ if and only if } \int_{0}^{\frac{1}{2}} \varkappa(ds)s < +\infty .$$

In one direction this follows since

$$\tfrac{1}{2}\,\eta_1'(t) = \tfrac{1}{2}\,\eta_1'(\tfrac{1}{2}) + \int_{\frac{1}{2}}^{t} \varkappa(ds)\;\eta_1(s)$$

$$\leq \tfrac{1}{2}\,\eta_1'(\tfrac{1}{2}) + \eta_1(t) \int_{\frac{1}{2}}^{t} \varkappa(ds)$$

and therefore

$$\eta_1'(t)/\eta_1(t) \leq c + \int_{\frac{1}{2}}^{t} \varkappa(ds)$$

$$\log \eta_1(t) \leq \log \eta_1(\tfrac{1}{2}) + c(t - \tfrac{1}{2}) + \int_{\frac{1}{2}}^{t} ds \int_{\frac{1}{2}}^{s} \varkappa(du).$$

In the other direction it follows from

$$\eta_1(t) = \eta_1(\tfrac{1}{2}) + (t - \tfrac{1}{2})\,\eta_1'(\tfrac{1}{2}) + \int_{\frac{1}{2}}^{t} ds \int_{\frac{1}{2}}^{s} \varkappa(du)\;\eta_1(u).$$

(These estimates are taken directly from Feller's paper [54].)

Define

$$\psi_0(x) = \eta_0(x)/\eta_0(0) \quad ; \quad \psi_1(x) = \eta_1(x)/\eta_1(0)$$

when they make sense. Otherwise let $\psi_0 \equiv 0$ or $\psi_1 \equiv 0$. For $u > 0$ let $\psi_{(u)}$ be a solution of

$$\tfrac{1}{2}\psi'_{(u)}(dx) = \psi_{(u)}(x)\,\kappa(dx) + u\psi_{(u)}(x)\,dx \tag{23.9}$$

which is positive and bounded away from 0 and let $\eta_{(u)0}, \eta_{(u)1}, \psi_{(u)0}, \psi_{(u)1}$ be defined in an analagous manner. Evidently

$$H\varphi(x) = \varphi(0)\psi_0(x) + \varphi(1)\,\psi_1(x)$$

$$H_u\varphi(x) = \varphi(0)\psi_{(u)0}(x) + \varphi(1)\psi_{(u)1}(x).$$

We begin by studying the Dirichlet spaces $(\underline{F}^{\sim}, E^{\sim})$ which are extensions of $(\underline{F}, E)$ and we assume at first that killing measure κ is bounded. Harmonic measure is always positive and finite at 0 and 1 and so there is no harm in working instead with the normalized measure ν^0 determined by

$$\nu^0(\{0\}) = \nu^0(\{1\}) = 1. \tag{23.10}$$

The calculation

$$\begin{aligned} U_{0,u}(0,1) &= u\int_0^1 dx\,\psi_0(x)\psi_{(u)1}(x) \\ &= \tfrac{1}{2}\int_0^1 \psi'_{(u)1}(dx)\,\psi_0(x) - \int \kappa(dx)\,\psi_0(x)\psi_{(u)1}(x) \\ &= \tfrac{1}{2}\int_0^1 \psi'_{(u)1}(dx)\,\psi_0(x) - \tfrac{1}{2}\int_0^1 \psi'_0(dx)\,\psi_{(u)1}(x) \end{aligned}$$

and an analagous one with the roles of 0 and 1 reversed establishes

(23.11) $$U_{0,u}(0,1) = -\tfrac{1}{2}\psi_0'(1) - \tfrac{1}{2}\psi_{(u)1}'(0) = \tfrac{1}{2}\psi_1'(0) + \tfrac{1}{2}\psi_{(u)0}'(1)\,.$$

After passage to the limit $u\uparrow\infty$

$$U_{0,\infty}(0,1) = -\tfrac{1}{2}\psi_0'(1) = \tfrac{1}{2}\psi_1'(0)$$

and therefore

$$N(\varphi,\varphi) = -\tfrac{1}{2}\psi_0'(1)\{\varphi(0)-\varphi(1)\}^2$$

$$+\int\varkappa(dx)\psi_0(x)\varphi^2(0) + \int\varkappa(dx)\psi_1(x)\varphi^2(1)\,.$$

Evidently condition (i) of Theorem 20.2 is satisfied for any choice of the subprobabilities $\mu(x,d\cdot)$. To simplify the notation we put

$$\mu_0(x) = \mu(x,\{0\});\ \mu_1(x) = \mu(x,\{1\})\ ;\ r(x) = 1-\mu_0(x)-\mu_1(x)\,.$$

In the notation of Section 20

(23.12) $$H^{\sim}\varphi(x) = \{\psi_0(x) + N(\mu_0\cdot\varkappa)(x)\}\varphi(0) + \{\psi_1(x) + \varkappa(\mu_1\cdot\varkappa)(x)\}\varphi(1)$$

$$\underline{\underline{N}}^{\sim} = \underline{\underline{N}}$$

$$N^{\sim}(\varphi,\varphi) = \{\textstyle\int\varkappa(dx)r(x)(\psi_0(x) + N\mu_0\cdot\varkappa(x))\}\varphi^2(0)$$

$$+\{\textstyle\int\varkappa(dx)r(x)(\psi_1(x) + N\mu_1\cdot\varkappa(x))\}\varphi^2(1)$$

$$+\{\tfrac{1}{2}\psi_1'(0) + \textstyle\int\varkappa(dx)\mu_0(x)\psi_1(x) + \int\varkappa(dx)\mu_1(x)\psi_0(x)$$

$$+\textstyle\int\varkappa(dx)\mu_0(x)N(\mu_1\varkappa)(x)\}\{\varphi(0)-\varphi(1)\}^2\,.$$

The possibilities for $\underline{\underline{H}}^{\sim}$ are still given by (22.7).

We look only at the case $\underline{\underline{H}}^{\sim} = \underline{\underline{N}}$. The most general $Q^{\sim}$ is

(23.13) $$Q^{\sim}(\varphi,\varphi) = N^{\sim}(\varphi,\varphi) + a\varphi^2(0) + b\varphi^2(1) + c\,\{\varphi(0) - \varphi(1)\}^2$$

with $a,b,c \geq 0$. Clearly $\underline{\underline{F}}^{\sim} = \underline{\underline{W}}$ and by the results in Section 20

(23.14) $$E^{\sim}(f,f) = W(f,f) + \int \varkappa(dx)\,(r(x)f^2(x) + \mu_0(x)\{f(x)-f(0)\}^2 + \mu_1(x)\{f(x)-f(1)\}^2) + af^2(0) + bf^2(1) + c\{f(0) - f(1)\}^2 .$$

A function f is in the domain of $A^{\sim}$ and $A^{\sim}f = g$ if and only if

(23.15) $$E^{\sim}(f,h) = - \int dx\,h(x)g(x)$$

for general h in $\underline{\underline{F}}^{\sim}$. The special case h in $\underline{\underline{F}}$ gives

(23.16) $$\tfrac{1}{2}f'(dx) - \varkappa(dx)\{f(x) - \mu_0(x)f(0) - \mu_1(x)f(1)\} = g(x)dx$$

and this must be supplemented by the boundary conditions

$$\mathscr{D}_n^{\sim} f(e_0) = af(0) + c\{f(0)-f(1)\} .$$

$$\mathscr{D}_n^{\sim} f(e_1) = bf(1) + c\{f(1) - f(0)\} .$$

A straightforward though slightly tedious calculation establishes

$$\mathscr{D}_n^{\sim} f(e_0) = \tfrac{1}{2}f'(0) + \int \varkappa(dx)\mu_0(x)\{f(x)-f(0)\}$$

$$\mathscr{D}_n^{\sim} f(e_1) = -\tfrac{1}{2}f'(1) + \int \varkappa(dx)\mu_1(x)\{f(x)-f(1)\} .$$

(We will give the details for an analagous computation in Section 24.) Thus f is in the domain of $A^{\sim}$ and $A^{\sim}f = g$ if and only if f satisfies (23.14) and also the two boundary conditions

$$(23.17)\quad \tfrac{1}{2}f'(0) + \int \varkappa(dx)\mu_0(x)\{f(x)-f(0)\} = af(0) + c\{f(0) - f(1)\}$$

$$-\tfrac{1}{2}f'(1) + \int \varkappa(dx)\mu_1(x)\{f(x)-f(1)\} = bf(1) + c\{f(1) - f(0)\} .$$

We continue now to assume that $\varkappa$ is bounded near 0 but we replace boundeness near 1 by a moment condition. Thus we assume

$$(23.18)\quad \int_0^{\frac{1}{2}} \varkappa(dx) < +\infty ; \quad \int_{\frac{1}{2}}^1 \varkappa(dx) = +\infty ; \quad \int_{\frac{1}{2}}^1 \varkappa(dx)(1-x) < +\infty .$$

The point of the moment condition is of course that ψ_1 is still nontrivial. Our classification results are essentially the same as before but our choices for μ_0 and μ_1 may impose restrictions. Also the boundary condition (23.17) at 1 must be interpreted in some formal sense. If $\int \varkappa(dx)r(x) = +\infty$ then every f in $\underline{\underline{F}}^{\sim}$ satisfies $f(1) = 0$ and the boundary point 1 plays no role. If $\int \varkappa(dx)r(x) < +\infty$ but $\int \varkappa(dx)\mu_0(x) = +\infty$ then every f in $\underline{\underline{F}}^{\sim}$ satisfies $f(0) = f(1)$ and the only nontrivial possibility is $\underline{\underline{N}}^{\sim} = \underline{\underline{N}}_p$. If $\int \varkappa(dx)r(x) < +\infty$ and $\int \varkappa(dx)\mu_0(x) < +\infty$ then $\underline{\underline{N}}^{\sim} = \underline{\underline{N}}$ and the classification is not really effected by the fact that $\int \varkappa(dx)\mu_1(x) = +\infty$. However the boundary condition (23.17) at 1 does not make sense in its present form. Every f in the domain of $A^{\sim}$ has a unique representation

$$f = f_o + f(0)\{\psi_0 + N\varkappa\mu_0\} + f(1)\{\psi_1 + N\varkappa\mu_1\}$$

with f_o in $\underline{\underline{F}}$ and indeed in the domain of A. We have $\tfrac{1}{2}f'_o(dx) = \varkappa(dx)f_o(x) + dx\,g(x)$ with g in $L^2(dx)$ and therefore in $L^1(dx)$. Every function in the domain of A is the difference of nonnegative functions in A and so by the third remark at the end of Section 20 the function f_o is in $L^1(\varkappa)$. Thus $f'_o(dx)$ has finite total variation and $f'_o(1)$ exists and it is still true that

$$\mathcal{D}_n^{\sim} f_o(e_1) = -\tfrac{1}{2} f_o'(1) + \int \varkappa(dx)\mu_1(x)\, f_o(x).$$

Also by convexity $\psi_0'(1)$, $N\mu_o \cdot \varkappa'(1)$ exist and again

$$\mathcal{D}_n^{\sim} \{\psi_0 + N\varkappa \cdot \mu_0\}'(e_1) = -\tfrac{1}{2}\psi_0'(1) - \tfrac{1}{2} N\varkappa \cdot \mu_0'(1)$$

$$+ \int \varkappa(dx)\mu_1(x)\{\psi_0(x) + N\varkappa \cdot \mu_0(x)\}\,.$$

However $\{\psi_1 + N\varkappa \cdot \mu_1\}'(1)$ does not exist and the appropriate integrals do not converge, although a computation establishes

$$\mathcal{D}_n^{\sim}\{\psi_1 + N\varkappa \cdot \mu_1\}'(e_1) = -\tfrac{1}{2}\psi_1'(0) - \int \varkappa(dx)(r(x) + \mu_0(x)\,\psi_1(x) \tag{23.19}$$

$$- \int \varkappa(dx)\mu_1(x)(\psi_0(x) + N\varkappa \cdot \mu_0(x) + N\varkappa \cdot r(x))$$

Thus the correct boundary condition at 1 is

$$-\tfrac{1}{2}\{f - f(1)\psi_1(x) - f(1)\, N\varkappa \cdot \mu_1(x)\}'(1) \tag{23.20}$$

$$+ \int \varkappa(dx)\mu_1(x)\{f(x) - f(1)\psi_1(x) - f(1) N\varkappa \cdot \mu_1(x)\}$$

$$+ f(1)\,\mathcal{D}_n^{\sim}\{\psi_1 + N\varkappa \cdot \mu_1\}(e_1)$$

$$= bf(1) + c\{f(1) - f(0)\}$$

which reduces to (23.17) when $\int \varkappa(dx) < +\infty$.

Finally we look briefly at the case when $\varkappa$ is unbounded and fails to satify a moment condition at both 0 and 1. Thus we assume

$$\int_0^{\frac{1}{2}} \varkappa(dx)\, x = +\infty \quad ; \quad \int_{\frac{1}{2}}^{1} \varkappa(dx)(1-x) = +\infty\,.$$

Then by (23.8) both ψ_0 and ψ_1 vanish identically and the manner in which Δ is topologically adjoined to $\underline{\underline{X}}$ is irrelevant. However condition (i) of Theorem 20.2

is violated unless $\mu_0 \cdot \varkappa$ and $\mu_1 \cdot \varkappa$ are both unbounded. By Theorem 20.8 the modified reflected Dirichlet form $E^{\sim}$ is given by (23.14) with a=b=c=0 and $\underline{\underline{F}}^{\sim}$ is the totality of f in W([0,1]) for which this expression converges. Clearly 0 and 1 "act like separate boundary points" if and only if $r \cdot \varkappa$ is bounded and $\mu_0 \cdot \varkappa$, $\mu_1 \cdot \varkappa$ are unbounded near different endpoints.

We turn now to the case when $P^{\sim}_t$ dominates P_t but $(\underline{\underline{F}}^{\sim}, E^{\sim})$ is not an extension of $(\underline{\underline{F}}, E)$. We will look only at the case when $\varkappa$ is bounded.

Let $\lambda(x, d\cdot)$, $x \in (0,1)$ be subprobabilities on (0,1) such that $\varkappa(dx)\lambda(x,dw)$ is symmetric and let $\rho(x) \geq 0$ satisfy $\lambda(x) + \rho(x) \leq 1$. It is clear that if $\varkappa$ is nondegenerate then there are many possible choices for λ and ρ. In the notation of Section 21, $\underline{\underline{F}}^{o} = \underline{\underline{F}}$ and

$$E^{o}(f,f) = \tfrac{1}{2}\int_0^1 dx\,\{f'(x)\}^2 + \int \varkappa(dx)\int \lambda(x,dw)\,\{f(x)-f(w)\}^2$$
$$+ \int \varkappa(dx)\rho(x) f^2(x)$$

We are interested in Dirichlet spaces $(\underline{\underline{F}}^{\sim}, E^{\sim})$ which are extensions of $(\underline{\underline{F}}^{o}, E^{o})$. Again there is no possible loss of generality in taking $\underline{X} \cup \Delta = [0,1]$.

A function f is in the domain of the local generator $\mathcal{O}^{o}$ associated with $(\underline{\underline{F}}^{o}, E^{o})$ if

(23.21) $\quad \tfrac{1}{2} f'(dx) = \varkappa(dx)\int \lambda(x,dw)\,\{f(x)-f(w)\} + \varkappa(dx)\rho(x) f(x) + dx\, g(x)$

with g in $L^2(dx)$ and then $\mathcal{O}^{o} f = g$. We first consider the case when the generator $A^{\sim}$ is contained in $\mathcal{O}^{o}$. The excursion space $\underline{\underline{N}}^{o} = \underline{\underline{N}}$. Let

(23.22) $\quad \psi^{o}_1(x) = \mathcal{P}_x\,(X_{\sigma(\Delta)} = 0)$

$\quad \psi^{o}_1(x) = \mathcal{P}_x\,(X_{\sigma(\Delta)} = 1)$.

It follows from the results in Section 14 that ψ_0^o, ψ_1^o belong to $\underline{\underline{F}}^{o,env}$ and that

(23.23) $$E^o(f, \psi_0^o) = E^o(f, \psi_1^o) = 0$$

for f in $\underline{\underline{F}}$ which is equivalent to the assertion that ψ_0^o, ψ_1^o satisfy the homogeneous equation

(23.23') $$\tfrac{1}{2}\psi^{o'}(dx) = \varkappa(dx)\int\lambda(x,dw)\{\psi^o(x)-\psi^o(w)\} + \varkappa(dx)\rho(x)\psi^o(x).$$

Similarly $\psi_{(u)0}^o$ and $\psi_{(u)1}^o$ defined by

(23.24) $$\psi_{(u)0}^o(x) = \mathcal{E}_x^o\, e^{-u\sigma(\Delta)}\, I(X_{\sigma(\Delta)} = 0)$$

$$\psi_{(u)1}^o(x) = \mathcal{E}_x^o\, e^{-u\sigma(\Delta)}\, I(X_{\sigma(\Delta)} = 1)$$

satisfy

(23.25) $$\tfrac{1}{2}\psi_{(u)}^{o'}(dx) = \varkappa(dx)\int\lambda(x,dw)\{\psi_{(u)}^o(x)-\psi_{(u)}^o(w)\} + \varkappa(dx)\rho(x)\psi_{(u)}^o(x) + u\,dx\,\psi_{(u)}^o(x).$$

Then

$$U_{0,u}^o(0,1) = u\int dx\,\psi_0^o(x)\,\psi_{(u)1}^o(x)$$

$$= \tfrac{1}{2}\int\psi_{(u)1}^{o'}(dx)\,\psi_0^o(x) - \int\varkappa(dx)\psi_0^o(x)\int\lambda(x,dw)\{\psi_{(u)1}^o(x)-\psi_{(u)1}^o(w)\}$$

$$- \int\varkappa(dx)\rho(x)\,\psi_0^o(x)\psi_{(u)1}^o(x)$$

$$= \tfrac{1}{2}\int\psi_{(u)1}^{o'}(dx)\psi_0^o(x) - \tfrac{1}{2}\int\psi_0^{o'}(dx)\,\psi_{(u)1}^o(x)$$

which together with a similar calculation reversing the roles of 0 and 1 establishes

(23.26) $$U_{0,u}^o(0,1) = -\tfrac{1}{2}\psi_0^{o'}(1) - \tfrac{1}{2}\psi'_{(u)1}(0) = \tfrac{1}{2}\psi_{(u)0}^{o'}(1) + \tfrac{1}{2}\psi_1^{o'}(0)$$

and after passage to the limit in u

(23.27) $$U^{o}_{0,\infty}(0,1) = -\tfrac{1}{2}\psi^{o'}_{0}(1) = \tfrac{1}{2}\psi^{o'}_{1}(0).$$

Thus

$$N^{o}(\varphi,\varphi) = \tfrac{1}{2}\psi^{o'}_{1}(0)\{\varphi(0)-\varphi(1)\}^{2} + \int \varkappa(dx)\rho(x)\psi^{o}_{0}(x)\varphi^{2}(0)$$

$$+ \int \varkappa(dx)\rho(x)\psi^{o}_{1}(x)\varphi^{2}(1).$$

Again the possibilities for $\underline{\underline{H}}^{\sim}$ are $\underline{\underline{N}}$, $\underline{\underline{N}}_0, \underline{\underline{N}}_1, \underline{\underline{N}}_p$. and the possibilities for $Q^{\sim}$ are the same as before except that N is replaced by N^{o}. Now

$$-\tfrac{1}{2}\int f'(dx)\psi^{o}_{0}(x) + \int \varkappa(dx)f(x)\int \lambda(x,dw)\{\psi^{o}_{0}(x) - \psi^{o}_{0}(w)\} + \int \varkappa(dx)\rho(x)f(x)\psi^{o}_{0}(x)$$

$$= -\tfrac{1}{2}\int f'(dx)\psi^{o}_{0}(x) + \tfrac{1}{2}\int \psi^{o'}_{0}(dx)f(x)$$

$$= \tfrac{1}{2}f'(0) + \tfrac{1}{2}\psi^{o'}_{0}(1)f(1) - \tfrac{1}{2}\psi^{o'}_{0}(0)f(0)$$

and

$$N^{o}(\gamma f,e_0) = -\tfrac{1}{2}\psi^{o'}_{0}(1)\{f(0)-f(1)\} + \int \varkappa(dx)\rho(x)\psi^{o}_{0}(x)f(0)$$

so that

$$\mathcal{D}^{o}_{n}f(e_0) = \tfrac{1}{2}f'(0) - \tfrac{1}{2}\psi^{o'}_{0}(0)f(0) + \tfrac{1}{2}\psi^{o'}_{0}(1)f(0)$$

$$+ \int \varkappa(dx)\rho(x)\psi^{o}_{0}(x)f(0)$$

$$= \tfrac{1}{2}f'(0)$$

since

$$\tfrac{1}{2}\int_0^1 \psi^{o'}_{0}(dx) - \int \varkappa(dx)\rho(x)\psi^{o}_{0}(x)$$

$$= \int \varkappa(dx)\int \lambda(x,dw)\{\psi^{o}_{0}(x) - \psi^{o}_{0}(w)\}$$

$$= 0$$

Thus the boundary conditions are also the same as before and in particular they are unaffected by λ.

The general case for $(F^{\sim}, E^{\sim})$ is easily disposed of. With μ_0, μ_1, r as before

$$H^{\sim}\varphi(x) = \{\psi_0^{o}(x) + N^{o}(\mu_0\rho\cdot\varkappa)(x)\}\varphi(0) + \{\psi_1^{o}(x) + N^{o}(\mu_1\rho\cdot\varkappa(x)\}\varphi(1)$$

$$\underline{\underline{N}}^{\sim} = \underline{\underline{N}}$$

$$N^{\sim}(\varphi,\varphi) = \{\int\varkappa(dx)\rho(x)r(x)(\psi_0^{o}(x) + N^{o}(\mu_0\rho\cdot\varkappa)(x))\}\varphi^2(0)$$

$$+ \{\int\varkappa(dx)\rho(x)r(x)(\psi_1^{o}(x) + N^{o}(\mu_1\rho\cdot\varkappa)(x))\}\varphi^2(1)$$

$$+ \{\tfrac{1}{2}\psi_1^{o'}(0) + \int\varkappa(dx)\rho(x)\mu_0(x)\psi_1^{o}(x) + \int\varkappa(dx)\rho(x)\mu_1(x)\psi_0^{o}$$

$$+ \int\varkappa(dx)\rho(x)\mu_0(x)N^{o}(\mu_1\rho\cdot\varkappa)(x)\}\{\varphi(0) - \varphi(1)\}^2$$

and now it turns out that

$$\mathscr{D}_n^{o\sim}f(e_0) = \tfrac{1}{2}f'(0) + \int\varkappa(dx)\rho(x)\mu_0(x)\{f(x) - f(0)\}$$

$$\mathscr{D}_n^{\sim}f(e_1) = -\tfrac{1}{2}f'(1) + \int\varkappa(dx)\rho(x)\mu_1(x)\{f(x) - f(1)\}\ .$$

24. Unbounded Scale.

Now let $W([0,\infty))$ be the collection of functions f defined and absolutely continuous on the closed half line $[0,\infty)$ such that the derivative f' is square integrable on $[0,\infty)$ and for f in $\underline{\underline{W}}([0,\infty))$ let

$$(24.1) \qquad W(f,f) = \tfrac{1}{2}\int_0^\infty dx \, \{f'(x)\}^2 .$$

Let $\underline{\underline{W}}_{abs}([0,\infty))$ be the subcollection of f in $\underline{\underline{W}}([0,\infty))$ satisfying the boundary condition

$$(24.2) \qquad f(0) = 0 .$$

Let κ be any nontrivial Radon measure on $(0,\infty)$ which is bounded near 0 and let m be any bounded measure on $(0,\infty)$ which charges every nonempty open subset of $(0,\infty)$. Throughout the section $\underline{\underline{X}} = (0,\infty)$, the role of the reference measure dx is played by m, the Dirichlet space $\underline{\underline{F}} = \underline{\underline{W}}_{abs}([0,\infty)) \cap L^2(\kappa) \cap L^2(m)$ and the Dirichlet form E is

$$E(f,f) = W(f,f) + \int \kappa(dx)\, f^2(x) .$$

Clearly $\underline{\underline{F}}_a^{env} = \underline{\underline{W}} \cap L^2(m)$ and $E^{res} = W$. There are many possible regularizing spaces for $\underline{\underline{F}}_a^{env}$ and so from the point of view of Theorem 20.3 there are many candidates for Δ. Any reasonable candidate will contain a point which we label 0 and which is adjoined to $\underline{\underline{X}}$ to form the closed half line $[0,\infty)$. If condition (i) of Theorem 20.2 is satisfied then (20.70) holds whenever φ in $\underline{\underline{N}}^\sim$ is nontrivial on $\Delta - \{0\}$. From the properties of the potential kernel $N(x,t)$ defined below it follows that any φ in $\underline{\underline{N}}^\sim$ must be constant on $\Delta - \{0\}$. Thus in studying extensions there is no real loss of generality in assuming that $\Delta = \{0,\infty\}$ with ∞ adjoined either as an isolated point or as the point at infinity. As in Section 23 let

24.2

$$\mu_0(x) = \mu(x,\{0\}) \; ; \; \mu_\infty(x) = \mu(x,\{\infty\}) \, .$$

We assume always that $\mu_0 \cdot \varkappa$ and $r \cdot \varkappa$ are bounded and that $\mu_\infty \cdot \varkappa$ is nontrivial. Condition (i) of Theorem 20.2 requires

$$\int \varkappa(dx)\mu_\infty(x) \, N\mu_\infty \cdot \varkappa(x) = +\infty \, . \tag{24.3}$$

In order to underline the significance of this condition we will also study the Dirichlet space $(\underline{\underline{F}}^\sim, E^\sim)$ obtained in the constructive part of Theorem 20.2 when (24.3) fails.

We assume first that (24.3) holds. Clearly $\underline{\underline{N}}^\sim$ is the full function space on Δ and

$$N^\sim(\varphi,\varphi) = \int \varkappa(dx) r(x) \{ \psi_0(x) + N\mu_0 \cdot \varkappa(x) \} \varphi^2(0)$$

$$+ \int \varkappa(dx) r(x) \, N\mu_\infty \cdot \varkappa(x) \, \varphi^2(\infty)$$

$$+ \{ \int \varkappa(dx)\mu_\infty(x) \, (\psi_0(x) + N\mu_0 \cdot \varkappa(x) \} \{\varphi(0) - \varphi(\infty)\}^2 .$$

Of course ψ_0 is defined as in Section 23. We will consider only the case $\underline{\underline{H}}^\sim = \underline{\underline{N}}^\sim$ and then the most general $Q^\sim$ is

$$Q^\sim(\varphi,\varphi) = N^\sim(\varphi,\varphi) + a\varphi^2(0) + b\varphi^2(\infty) + c\{\varphi(0) - \varphi(\infty)\}^2 .$$

It follows from Theorem 20.2 that every f in $\underline{\underline{F}}^\sim$ has a unique representation

$$f = f_o + \gamma f(0) \{ \psi_0 + N\mu_0 \cdot \varkappa \} + \gamma f(\infty) \, N\mu_\infty \cdot \varkappa \tag{24.4}$$

with f_o in $\underline{\underline{F}}$ and that

$$E^\sim(f,f) = W(f,f) + \int \varkappa(dx) r(x) f^2(x) \tag{24.5}$$

$$+ \int \varkappa(dx)\mu_0(x) \{f(x) - \gamma f(0)\}^2 + \int \varkappa(dx)\mu_\infty(x) \{f(x) - \gamma f(\infty)\}^2 .$$

Moreover by Theorem 20.8 $\gamma f(0) = f(0)$ and f in $\underline{\underline{W}}([0,\infty))$ belongs to $\underline{\underline{F}}^{\sim}$ if and only if there exists $\gamma f(\infty)$ such that (24.5) converges and then $\gamma f(\infty)$ is unique. It is true that if f is in domain $A^{\sim}$ then also $\gamma f(\infty) = \operatorname{Lim}_{x\uparrow\infty} f(x)$, but this need not be the case for general f in $\underline{\underline{F}}^{\sim}$. As in Section 23 every f in the domain of $A^{\sim}$ satisfies

$$(24.6) \qquad \tfrac{1}{2} f'(dx) - \varkappa(dx)\{f(x) - \mu_0(x)f(0) - \mu_\infty(x)\gamma f(\infty)\} = g(x)m(dx)$$

with g in $L^2(m)$ and this must be supplemented by the two boundary conditions

$$(24.7) \qquad \mathcal{D}_n^{\sim} f(e_0) = af(0) + c\{f(0) - \gamma f(\infty)\}$$

$$\mathcal{D}_n^{\sim} f(e_\infty) = b\gamma f(\infty) + c\{\gamma f(\infty) - f(0)\}.$$

We give the details now for identifying the left hand sides.

First we establish

$$(24.8) \qquad -\tfrac{1}{2}\int f'(dx) N\mu_0 \cdot \varkappa(x) + \int \varkappa(dx) f(x) N\mu_0 \cdot \varkappa(x)$$

$$= \int \varkappa(dx)\mu_0(x) f(x) - \int \varkappa(dx)\mu_0(x)\psi_0(x) f(0).$$

It can be verified with the help of the third remark at the end of Section 20 that individual terms converge separately for f in the domain of $A^{\sim}$. In verifying (24.8) we must use the fact that the potential operator N has the density

$$(24.9) \qquad N(x,t) = (2/\eta_\infty'(0))\, \eta_\infty(x \wedge t)\, \psi_0(x \vee t)$$

where η_∞ is defined as in Section 24 and where $x \wedge t = \min(x,t)$ and $x \vee t = \max(x,t)$. Then the left side of (24.8)

$$(24.10) \qquad = (2/\eta_\infty'(0)) \int_0^\infty \{-\tfrac{1}{2} f'(dx)\psi_0(x) + \varkappa(dx) f(x)\psi_0(x)\} \int_0^x \varkappa(dt)\mu_0(t)\eta_\infty(t)$$

$$+ (2/\eta_\infty'(0)) \int_0^\infty \{-\tfrac{1}{2} f'(dx)\eta_\infty(x) + \varkappa(dx) f(x)\eta_\infty(x)\} \int_x^\infty \varkappa(dt)\mu_0(t)\psi_0(t)$$

$$=(2/\eta_\infty'(0))\int_0^\infty \varkappa(dt)\mu_0(t)\eta_\infty(t)\int_t^\infty\{-\tfrac{1}{2}f'(dx)\psi_0(x)+\tfrac{1}{2}\psi_0'(dx)f(x)\}$$

$$+(2/\eta_\infty'(0))\int_0^\infty \varkappa(dt)\mu_0(t)\psi_0(x)\int_0^t\{-\tfrac{1}{2}f'(dx)\eta_\infty(x)+\tfrac{1}{2}\eta_\infty'(dx)f(x)\}$$

and (24.8) follows since $\psi_0(t)\eta_\infty'(t)-\psi_0'(t)\eta_\infty(t)=\eta_\infty'(0)$ and since the appropriate boundary terms at infinity vanish. Thus

$$\mathcal{D}_n^\sim f(e_0) = -N^\sim(\gamma f,e_0)-\tfrac{1}{2}\int f'(dx)(\psi_0(x)+N\mu_0\cdot\varkappa(x))$$

$$+\int\varkappa(dx)\{f(x)-\mu_0(x)f(0)-\mu_\infty(x)\gamma f(\infty)\}\{\psi_0(x)+N\mu_0\cdot\varkappa(x)\}$$

$$=-f(0)\{\int\varkappa(dx)r(x)\psi_0(x)+\int\varkappa(dx)r(x)\,N\mu_0\cdot\varkappa(x)\}$$

$$+\{\gamma f(\infty)-f(0)\{\int\varkappa(dx)\mu_\infty(x)\psi_0(x)+\int\varkappa(dx)\mu_\infty(x)N\mu_0\cdot\varkappa(x)\}$$

$$+\tfrac{1}{2}f'(0)-\tfrac{1}{2}f(0)\psi_0'(0)+\int\varkappa(dx)\mu_0(x)f(x)-\int\varkappa(dx)\mu_0(x)\psi_0(x)f(0)$$

$$-f(0)\int\varkappa(dx)\mu_0(x)\{\psi_0(x)+N\mu_0\cdot\varkappa(x)\}$$

$$-\gamma f(\infty)\int\varkappa(dx)\mu_\infty(x)\{\psi_0(x)+N\mu_0\cdot\varkappa(x)\}$$

and after some cancellation

(24.11) $$\mathcal{D}_n^\sim f(c_0)=\tfrac{1}{2}f'(0)+\int\varkappa(dx)\mu_0(x)\{f(x)-f(0)\}\,.$$

For $\mathcal{D}_n^\sim f(e_\infty)$ we must take more care since individual terms do not converge. We consider separately the cases $f=f_0+f(0)\{\psi_0+N\mu_0\cdot\varkappa\}$ and $f=N\mu_1\cdot\varkappa$. In the former case individual terms still converge and essentially the same calculations as above establish

$$\mathcal{D}_n^\sim f(e_\infty)=\int_0^\infty\varkappa(dt)\mu_\infty(t)f(t).$$

In the latter case put $f=N\mu_\infty\cdot\varkappa$ and $g=1-f$. Then

$$\mathcal{A}^{\sim}_{n} f(e_\infty) = - N^{\sim}(e_\infty, e_\infty) - \int \{ \tfrac{1}{2} f'(dx) - \varkappa(dx) f(x) + \varkappa(dx)\mu_\infty(x) \} f(x)$$

$$= - \int \varkappa(dx) r(x) N\mu_\infty \cdot \varkappa(x) - \int \varkappa(dx)\mu_\infty(x) (\psi_0(x) + N\mu_0 \cdot \varkappa(x))$$

$$+ \tfrac{1}{2} \int g'(dx) f(x) - \int \varkappa(dx) g(x) f(x) + \int \varkappa(dx) (r(x) + \mu_0(x)) f(x)$$

$$= \tfrac{1}{2} \int g'(dx) f(x) - \int \varkappa(dx) g(x) f(x) - \int \varkappa(dx) \mu_\infty(x) \psi_0(x)$$

A computation analagous to (24.10) shows that

$$\tfrac{1}{2} \int g'(dx) f(x) - \int \varkappa(dx) g(x) f(x)$$

$$= - \int \varkappa(dx) \mu_\infty(x) g(x) + \int \varkappa(dx) \mu_\infty(x) \psi_0(x)$$

and we conclude that for general f in the domain of $A^{\sim}$

(24.12) $$\mathcal{A}^{\sim}_{n} f(e_\infty) = \int \varkappa(dx) \mu_\infty(x) \{(x) - \gamma f(\infty)\} .$$

We finish by considering the case when $\varkappa$ is bounded and therefore (24.3) fails. Let $\underline{\underline{N}}^{\sim}$ and $\underline{\underline{G}}^{\sim}$ be as before, let $G^{\sim}_{u}$ be the resolvent constructed in the proof of Theorem 20.2 and let $(\underline{\underline{F}}^{\sim}, E^{\sim})$ be the associated Dirichlet space. This makes sense even though condition (i) is violated. It is clear that $\underline{\underline{F}}$ has codimension one in $\underline{\underline{F}}^{\sim}$ and indeed that ψ_0 generates a complementary subspace for $\underline{\underline{F}}$. We will prove that

(24.13) $$E^{\sim}(f,f) = W(f,f) + \int \varkappa(dx) r(x) f^2(x) + \int \varkappa(dx) \mu_0(x) \{f(x) - f(0)\}^2$$

$$+ \{a + (bc)/(b + c + \int \varkappa(dx) \mu_\infty(x))\} f^2(0)$$

$$+ \tfrac{1}{2} \{b + c + \int \varkappa(dx) \mu_\infty(x)\}^{-1} \int \varkappa(dx) \mu_\infty(x) \int \varkappa(dw) \mu_\infty(w) \{f(x) - f(w)\}^2$$

$$+ (b / \{b + c + \int \varkappa(dx) \mu_\infty(x)\}) \int \varkappa(dx) \mu_\infty(x) f^2(x)$$

$$+ (c / \{b + c + \int \varkappa(dx) \mu_\infty(x)\}) \int \varkappa(dx) \mu_\infty(x) \{f(x) - f(0)\}^2 .$$

The significant feature is that $(\underline{\underline{F}}^{\sim}, E^{\sim})$ is not an extension of $(\underline{\underline{F}}, E)$. The point ∞ is "not really there." It might be appropriate to call ∞ an "illusory point."

In proving (24.13) we consider first the case $f \in \underline{\underline{F}}$ and then we can assume $f = G_1 g$ with g in $L^2(m)$. By the proof of Theorem 20.2

$$E_1^{\sim}(H_1^{\sim} R_{(1)}^{\sim} \pi_1^{\sim} g, G_1 g)$$

$$= \operatorname{Lim}_{v \uparrow \infty} \int m(dx)\, H_1^{\sim} R_{1+v}^{\sim} \pi_1^{\sim} g(x) g(x).$$

Again to simplify the notation we replace harmonic measure ν by the normalized measure ν^{o} determined by

$$\nu^{o}(\{0\}) = \nu^{o}(\{\infty\}) = 1.$$

Let $Q_{(v)}^{\sim}(0,0)$, $Q_{(v)}^{\sim}(0,\infty)$, $Q_{(v)}^{\sim}(\infty,\infty)$ be defined by

$$Q_{(v)}^{\sim}(\varphi,\varphi) = Q_{(v)}^{\sim}(0,0)\,\varphi^2(0) + Q_{(v)}^{\sim}(\infty,\infty)\,\varphi^2(\infty) \tag{24.14}$$

$$+ 2\, Q_{(v)}^{\sim}(0,\infty)\,\varphi(0)\varphi(\infty).$$

It is easy to see that $\operatorname{Lim}_{v \uparrow \infty} Q_{(v)}^{\sim}(0,0) = +\infty$ while $Q_{(v)}^{\sim}(0,\infty)$, $Q_{(v)}^{\sim}(\infty,\infty)$ have finite limits and since $R_{(v)}^{\sim}$ inverts $Q_{(v)}^{\sim}$ it follows that

$$\operatorname{Lim}_{v \uparrow \infty} R_{(v)}^{\sim} e_0 = 0$$

$$\operatorname{Lim}_{v \uparrow \infty} R_{(v)}^{\sim} e_{\infty} = \operatorname{Lim}_{v \uparrow \infty} (1/Q_{(v)}^{\sim}(\infty,\infty))\, e_{\infty}.$$

$$= \{ b+c+\int \varkappa(dx)\mu_{\infty}(x) \}^{-1} e_{\infty}.$$

Thus

$$E_1^{\sim}(H_1^{\sim} R_{(1)}^{\sim} \pi_1^{\sim} g,\ G_1 g)$$

$$= \{b+c+\int \varkappa(dx)\mu_{\infty}(x)\}^{-1} \{\int \varkappa(dx)\mu_{\infty}(x) f(x)\}^2$$

$$E_1^\sim(f,f) = E_1^\sim(G_1^\sim g, G_1 g) - E_1^\sim(H_1^\sim R_{(1)}^\sim \pi_1^\sim g, G_1 g)$$

$$= E_1(G_1 g, G_1 g) - E_1^\sim(H_1^\sim R_{(1)}^\sim \pi_1^\sim g, G_1 g)$$

$$E^\sim(f,f) = E(f,f) - \{b+c+\int \varkappa(dx)\mu_\infty(x)\}^{-1}\{\int \varkappa(dx)\mu_\infty(x)f(x)\}^2$$

which is equivalent to (24.13)

We still must verify (24.13) for $E^\sim(\psi_0,\psi_0)$ and for $E^\sim(\psi_0,f)$ with $f \in \underline{\underline{F}}$. Our proofs in these cases depend on some complicated although pedestrian computations. To save space we give these computations with certain negligible terms omitted from the very beginning. The notation (24.14) is taken for granted throughout. Also we again introduce special notations such as

$$\varkappa r \psi_0 = \int \varkappa(dx) r(x) \psi_0(x) \tag{24.15}$$

$$\varkappa \mu_0 N\mu_0 \varkappa = \int \varkappa(dx)\mu_0(x)\, N\mu_0 \cdot \varkappa(x).$$

Then

$$Q^\sim(0,0) = \varkappa r \psi_0 + \varkappa r N\mu_0 \varkappa + \varkappa\mu_\infty \psi_0 + \varkappa\mu_\infty N\mu_0 \varkappa + a + c$$

$$U^\sim_{0,u}(0,0) \sim -\tfrac{1}{2}\psi'_{(u)0}(0) + \tfrac{1}{2}\psi'_0(0) + 2\varkappa\mu_0\psi_0 + \varkappa\mu_0 N\mu_0 \varkappa$$

$$Q^\sim_{(u)}(0,0) \sim -\tfrac{1}{2}\psi'_{(u)0}(0) + \varkappa\mu_0 + a + c$$

$$Q^\sim(0,\infty) = -\varkappa\mu_\infty\psi_0 - \varkappa\mu_\infty N\mu_\infty \varkappa - c$$

$$U^\sim_{0,u}(0,\infty) \sim \varkappa\mu_\infty\psi_0 + \varkappa\mu_\infty N\mu_0 \varkappa$$

$$Q^\sim_{(u)}(0,\infty) \sim -c$$

$$Q^\sim(\infty,\infty) = \varkappa r N\mu_\infty \varkappa + \varkappa\mu_\infty\psi_0 + \varkappa\mu_\infty N\mu_0 \varkappa + b + c$$

$$U^{\sim}_{0,u}(\infty,\infty) \sim \varkappa\mu_\infty N\mu_\infty \varkappa$$

$$Q^{\sim}_{(u)}(\infty,\infty) \sim \varkappa\mu_\infty + b + c$$

$$\det . Q^{\sim}_{(u)} \sim -\tfrac{1}{2}\psi'_{(u)0}(0)\{\varkappa\mu_\infty + b + c\} + \{\varkappa\mu_0 + a + c\}\{\varkappa\mu_\infty + b + c\} - c^2$$

$$\{\det Q^{\sim}_{(u)}\}^{-1} \sim (-\tfrac{1}{2}\psi'_{(u)0}(0)\{\varkappa\mu_\infty + b + c\})^{-1}\,[1-$$

$$(\varkappa\mu_0 + a + c)/(-\tfrac{1}{2}\psi'_{(u)0}(0)) + c^2/(-\tfrac{1}{2}\psi'_{(u)0}(0)\{\varkappa\mu_\infty + b + c\})\,]$$

$$R^{\sim}_{(u)}(0,0) = Q^{\sim}_{(u)}(\infty,\infty)/(\det Q^{\sim}_{(u)}) \sim (\varkappa\mu_\infty + b + c)/(\det Q^{\sim}_{(u)})$$

$$R^{\sim}_{(u)}(\infty,\infty) = Q^{\sim}_{(u)}(0,0)/(\det Q^{\sim}_{(u)}) \sim (-\tfrac{1}{2}\psi'_{(u)0}(0) + \varkappa\mu_0 + a + c)/(\det Q^{\sim}_{(u)})$$

$$R^{\sim}_{(u)}(0,\infty) = -Q^{\sim}_{(u)}(0,\infty)/(\det Q^{\sim}_{(u)}) \sim c/(\det Q^{\sim}_{(u)})$$

$$u\pi^{\sim}_u\psi_0 \sim (-\tfrac{1}{2}\psi'_{(u)0}(0) + \tfrac{1}{2}\psi'_0(0) + \varkappa\mu_0\psi_0)e_0 + \varkappa\mu_\infty\psi_0 e_\infty$$

$$u^2\int \nu^0(dy)\pi^{\sim}_u\psi_0(y)R^{\sim}_{(u)}\pi^{\sim}_u\psi_0(y)$$

$$\sim \{\det Q^{\sim}_{(u)}\}^{-1}\{-\tfrac{1}{2}\psi'_{(u)0}(0) + \tfrac{1}{2}\psi'_0(0) + \varkappa\mu_0\psi_0\}^2\{\varkappa\mu_\infty + b + c\}$$

$$+ \{\det Q^{\sim}_{(u)}\}^{-1}\{\varkappa\mu_\infty\psi_0\}^2\{-\tfrac{1}{2}\psi'_{(u)}(0) + \varkappa\mu_0 + a + c\}$$

$$+ 2\{\det Q^{\sim}_{(u)}\}^{-1}c\{-\tfrac{1}{2}\psi'_{(u)0}(0) + \tfrac{1}{2}\psi'_0(0) + \varkappa\mu_0\psi_0\}\{\varkappa\mu_\infty\psi_0\}$$

$$\sim -\tfrac{1}{2}\psi'_{(u)0}(0) + 2\{\tfrac{1}{2}\psi'_0(0) + \varkappa\mu_0\psi_0\}$$

$$+ \{\varkappa\mu_\infty\psi_0\}^2/\{\varkappa\mu_\infty + b + c\}$$

$$+ 2c\{\varkappa\mu_\infty\psi_0\}/\{\varkappa\mu_\infty + b + c\}$$

$$-(\varkappa\mu_0 + a + c) + c^2/\{\varkappa\mu_\infty + b + c\}$$

$$E^{\sim}(\psi_0,\psi_0) \sim u \int m(dx)\,\{\psi_0(x) - uG_u\,\psi_0(x)\}\,\psi_0(x)$$

$$- u^2\int m(dx) H_u^{\sim} R^{\sim}_{(u)} \Pi^{\sim}_u \psi_0(x)\psi_0(x)$$

$$\sim - u\int m(dx)\psi_0(x)\psi_{(u)0}(x) - u^2\int \nu^0(dy)\Pi^{\sim}_u \psi_0(y) R^{\sim}_{(u)} \Pi^{\sim}_u \psi_0(y)$$

$$\sim - \tfrac{1}{2}\psi'_{(u)0}(0) + \tfrac{1}{2}\psi'_0(0) - u^2\int \nu^0(dy)\Pi^{\sim}_u \psi_0(y) R^{\sim}_{(u)} \Pi^{\sim}_u \psi_0(y)$$

$$= - \psi'_0(0) - 2\varkappa\mu_0\psi_0 - (\varkappa\mu_\infty\psi_0)^2/(\varkappa\mu_\infty + b + c)$$

$$- 2c(\varkappa\mu_\infty\psi_0)/(\varkappa\mu_\infty + b + c)$$

$$+ \varkappa\mu_0 + a + c - c^2/(\varkappa\mu_\infty + b + c \quad .$$

But this is consistent with (24.13) since

$$W(\psi_0,\psi_0) + \int\varkappa(dx)r(x)\psi_0^2(x) + \int\varkappa(dx)\mu_0(x)\{\psi_0(x) - 1\}^2$$

$$+ a + bc/(b+c+ \varkappa\mu_\infty)$$

$$+ \{(\varkappa\mu_\infty)/(b+c+\varkappa\mu_\infty)\} \int\varkappa(dx)\mu_\infty(x)\psi_0^2(x)$$

$$- \{b+c+\varkappa\mu_\infty\}^{-1}\{\int\varkappa(dx)\mu_\infty(x)\psi_0(x)\}^2$$

$$+ \{b/(b+c+\varkappa\mu_\infty)\}\int\varkappa(dx)\mu_\infty(x)\psi_0^2(x)$$

$$+ \{c/(b+c+\varkappa\mu_\infty)\}\int\varkappa(dx)\mu_\infty(x)\{\psi_0(x)-1\}^2$$

$$= W(\psi_0,\psi_0) + \varkappa\psi_0^2 - 2\varkappa\mu_0\psi_0 + \varkappa\mu_0$$

$$+ a + (b c)/(b+c+\varkappa\mu_\infty)$$

$$- 2c(\varkappa\mu_\infty\psi_0)/(b+c+\varkappa\mu_\infty)$$

$$+ (c\varkappa\mu_\infty)/(b+c+\varkappa\mu_\infty)$$

$$- (\varkappa\mu_\infty\psi_0)^2/(b+c+\varkappa\mu_\infty)$$

$$= \tfrac{1}{2}\psi_0'(0) - 2\varkappa\mu_0\psi_0 + \varkappa\mu_0 - 2c(\varkappa\mu_\infty\psi_0)/(b+c+\varkappa\mu_\infty)$$

$$- (\varkappa\mu_\infty\psi_0)^2/(b+c+\varkappa\mu_\infty) + a + c - c^2/(b+c+\varkappa\mu_\infty).$$

In checking the cross term $E^\sim(\psi_0, f)$ we can assume that $f \in C_{com}(\underline{\underline{X}})$. The point is that

$$u\int m(dx)\psi_{(u)0}(x)f(x) = \tfrac{1}{2}\int \psi'_{(u)0}(dx)f(x) - \int \varkappa(dx)\,\psi_{(u)0}(x)f(x)$$

is bounded independent of u since for $\varepsilon > 0$

$$-\tfrac{1}{2}\psi'_{(u)0}(\varepsilon) \leq 1/\varepsilon \ .$$

Thus

$$u\widetilde{\pi}_u f \sim \{u\int m(dx)f(x)\psi_{(u)0}(x) + \varkappa\mu_0 f\}\, e_0 + \{\varkappa\mu_\infty f\}\, e_\infty$$

$$E^\sim(f,f) \sim u\int m(dx)f(x)\psi_{(u)0}(x) - u^2\int \nu^0(dy)\widetilde{\pi}_u f(y) R^\sim_{(u)}\widetilde{\pi}_u\psi_0(y)$$

$$\sim u\int m(dx)f(x)\psi_{(u)0}(x)$$

$$-(\det Q^\sim_{(u)})^{-1}(u\int m(dx)f(x)\psi_{(u)0}(x)+\varkappa\mu_0 f)(-\tfrac{1}{2}\psi'_{(u)0}(0)+\tfrac{1}{2}\psi_0'(0) + \varkappa\mu_0\psi_0)(b+c+\varkappa\mu_\infty)$$

$$-(\det Q^\sim_{(u)})^{-1}(\varkappa\mu_\infty f)(\varkappa\mu_\infty\psi_0)(-\tfrac{1}{2}\psi'_{(u)0}(0) + \varkappa\mu_0 + a + c)$$

$$-(\det Q^\sim_{(u)})^{-1}c(u\int m(dx)f(x)\psi_{(u)0}(x) + \varkappa\mu_0 f)(\varkappa\mu_\infty\psi_0)$$

$$-(\det Q^\sim_{(u)})^{-1}c(\varkappa\mu_\infty f)(-\tfrac{1}{2}\psi'_{(u)0}(0) + \tfrac{1}{2}\psi_0'(0) + \varkappa\mu_0\psi_0)$$

$$\sim u\int m(dx)f(x)\psi_{(u)0}(x)$$

$$- u\int m(dx)f(x)\psi_{(u)0}(x) - \varkappa\mu_0 f$$

$$-(\varkappa\mu_\infty f)(\varkappa\mu_\infty\psi_0)/(b+c+\varkappa\mu_\infty)$$

$$-\ c(\varkappa\mu_\infty f)/(b+c+\varkappa\mu_\infty)$$

$$=-\varkappa\mu_0 f-(\varkappa\mu_\infty f)(c+\varkappa\mu_\infty\psi_0)/(b+c+\varkappa\mu_\infty)$$

which is consistent with (24.13) since

$$W(f,\psi_0)+\int\varkappa(dx)r(x)f(x)\psi_0(x)+\int\varkappa(dx)\mu_0(x)f(x)\{\psi_0(x)-1\}$$

$$+\tfrac{1}{2}\{b+c+\varkappa\mu_\infty\}^{-1}\int\varkappa(dx)\mu_\infty(x)\int\varkappa(dw)\mu_\infty(w)\{f(x)-f(w)\}\{\psi_0(x)-\psi_0(w)\}$$

$$+\{b/(b+c+\varkappa\mu_\infty)\}\varkappa\mu_\infty f\psi_0$$

$$+\{c/(b+c+\varkappa\mu_\infty)\}\int\varkappa(dx)\mu_\infty(x)f(x)\{\psi_0(x)-1\}$$

$$=W(f,\psi_0)+\int\varkappa(dx)f(x)\psi_0(x)-\varkappa\mu_0 f$$

$$-(\varkappa\mu_\infty f)(\varkappa\mu_\infty\psi_0)/(b+c+\varkappa\mu_\infty)-\{c/(b+c+\varkappa\mu_\infty)\}\varkappa\mu_\infty f$$

and since by the results in Section 14

$$W(f,\psi_0)+\int\varkappa(dx)f(x)\psi_0(x)=0.$$

25. Infinitely Divisible Processes

In this section we look briefly at the case when $(\underline{F}, E)$ is the absorbed process relative to (0,1) for a symmetric infinitely divisible process. We do not attempt to give a complete treatment.

First some notations. Let $\underline{R}$ be the real line and let dx be normalized Lebesgue measure on $\underline{R}$. Let $\underline{W}$ be the collection of absolutely continuous f on $\underline{R}$ such that the derivative $f' \in L^2(\underline{R})$ and for $f \in \underline{W}$ define

$$W(f,f) = \tfrac{1}{2}\int dx\, \{f'(x)\}^2 .$$

Also let $\underline{W}_a = \underline{W} \cap L^2(\underline{R})$. Let $\ell(dx)$ be a Radon measure on the deleted real line $\underline{R} - \{0\}$ satisfying the symmetry condition

$$(25.1) \qquad \int \ell(dx)\varphi(x) = \int \ell(dx)\varphi(-x)$$

and the integrability condition

$$(25.2) \qquad \int_0^1 \ell(dx)x^2 + \int_1^\infty \ell(dx) < +\infty .$$

Define

$$(25.3) \qquad L(f,f) = \tfrac{1}{2}\int dx \int \ell(dw)\, \{f(x) - f(x+w)\}^2$$

and let $\underline{L}_a$ be the set of $f \in L^2(\underline{R})$ for which (25.3) converges. With Fourier transforms it is easy to check that

$$(25.4) \qquad \underline{W}_a \subset \underline{L}_a .$$

The pair (W_a, W) is easily seen to be a regular Dirichlet space on $\underline{R}$; the associated process is "standard Brownian motion." From this and (25.4) it follows that also $(\underline{W}_a, W+L)$ is a regular Dirichlet space on

$L^2(\underline{\underline{R}})$. Also (25.4) together with an approximation argument using convolutions guarantees that the pair $(\underline{\underline{L}}_a, L)$ is a regular Dirichlet space on $L^2(\underline{\underline{R}})$. It is well known that these are the most general translation invariant Dirichlet spaces on $L^2(\underline{\underline{R}})$.

For $f \in \underline{\underline{W}}_a$ the relation

$$f(x) = \int_0^1 dy\, f(y) + \int_0^1 dy \int_y^x dw\, f'(w)$$

yields the crude estimate

$$|f(x)|^2 \leq 2\int dy\, f^2(y) + (2 + 2(x)^2)\, W(f,f)$$

which is enough to guarantee that the singletion $\{x\}$ is nonpolar for $(\underline{\underline{W}}_a, W)$ and therefore also for $(\underline{\underline{W}}_a, W + L)$. Singletons may or may not be polar for $(\underline{\underline{L}}_a, L)$. (See [52] or [54] for appropriate criteria.) Our general results in Section 4 only guarantee the existence of probabilities $\mathcal{P}_x$ for x outside some polar set. However it is easy to use translation invariance to establish the existence of $\mathcal{P}_x$ for all x in such a way that

$$\mathcal{E}_x\, \varphi(X_t) = \mathcal{E}_0\, \varphi(x + X_t)$$

and we take this for granted below.

Before going on to discuss the absorbed spaces, we prove the elementary

<u>Theorem 25.1</u>. (i) <u>The Dirichlet spaces</u> $(\underline{\underline{W}}_a, W)$ <u>and</u> $(\underline{\underline{W}}_a, W+L)$ <u>are always irreducible</u>.

(ii) <u>The Dirichlet space</u> $(\underline{\underline{L}}_a, L)$ <u>is irreducible unless</u> ℓ <u>is concentrated on a lattice</u>. ///

Proof. Conclusion (i) follows since singletons are nonpolar and the function

$$h_1(x) = \mathscr{E}_0 e^{-\sigma(\{x\})}$$

is quasi-continuous and therefore continuous (Theorem 4.14-(iii)). For (ii) let A be a proper invariant set and note that by Lemma 1.1

$$\int dx\, 1_A(x) f(x) \int \ell(dy) 1_{A^c}(x+g) g(y) = 0$$

for f,g in $\underline{\underline{L}}_a$ such that $fg = 0$ almost everywhere. In particular this is true for f,g in $\underline{\underline{W}}_a$ with disjoint supports and so by routine measure theoretic arguments

(25.5) $$\int \ell(dy) \{ \int dx\, 1_A(x)\, 1_{A^c}(x+y) \} = 0.$$

Thus ℓ is concentrated on the set of y satisfying

(25.6) $$\int dx\, 1_A(x)\, 1_{A^c}(x+y) = \int dx\, 1_A(x)\, 1_{A^c}(x-y) = 0.$$

But for y satisfying (25.6) and for φ bounded and integrable

$$\int dx\, 1_A(x)\, \varphi(x+y)$$

$$= \int dx\, 1_A(x)\, 1_A(x+y) \varphi(x+y)$$

$$= \int dx\, 1_A(x) \varphi(x)\, 1_A(x-y)$$

$$= \int dx\, 1_A(x)\, \varphi(x).$$

Thus the set of y satisfying (25.6) forms a closed subgroup of $\underline{R}$. Since $1_A(x) \cdot dx$ is a nontrivial Radon measure which is not Lebesgue measure, (25.6) cannot be valid for all y and the theorem follows. ///

Now let $\underline{\underline{X}} = (0,1)$ and let $(\underline{\underline{F}}, E)$ be the absorbed space relative to $\underline{\underline{X}}$ either for $(\underline{\underline{W}}_a, W+L)$ or for $(\underline{\underline{L}}_a, L)$. For $0 < x < 1$ let

$$\kappa(x) = \int_x^\infty \ell(dy) + \int_{1-x}^\infty \ell(dy).$$

If $(\underline{\underline{F}}, E)$ is the absorbed space for $(\underline{\underline{W}}_a, W+L)$, then

$$\underline{\underline{F}} = \underline{\underline{W}}_{abs}([0,1])$$

$$E(f,f) = W(f,f) + \tfrac{1}{2} \int_0^1 dx \int_{-x}^{1-x} \ell(dy) \{f(x)-f(x+y)\}^2 . \tag{25.7}$$

$$+ \int_0^1 dx\, \kappa(x) f^2(x).$$

If $(\underline{\underline{F}}, E)$ is the absorbed space for $(\underline{\underline{L}}_a, L)$ then

$$E(f,f) = \tfrac{1}{2} \int_0^1 dx \int_{-x}^{1-x} \ell(dy) \{f(x)-f(y)\}^2 + \int_0^1 dx\, \kappa(x) f^2(x) \tag{25.8}$$

and $\underline{\underline{F}}$ is (possibly) a refinement of the set of $f \in L^2(\underline{\underline{X}}, dx)$ for which (25.8) converges. (In some of the above integrals we are ignoring endpoints which is not really legitimate if ℓ has atoms. However it is easy to check that the resulting ambiguities are harmless.)

The classification of extensions differs according to whether or not ℓ satisfies the first moment condition

$$\int_0^1 \ell(dy) y < +\infty \tag{25.9}$$

since the relation

$$\int_0^1 dx \int_{1-x}^{\infty} \ell(dy) = \int_0^1 dx \int_x^{\infty} \ell(dy) = \int_0^{\infty} \ell(dy)\ (y \wedge 1)$$

shows that this condition is equivalent to

$$\int_0^1 dx\, \varkappa(x) < +\infty \ . \tag{25.9'}$$

We look first at the case when $(\underline{\underline{F}}, E)$ is the absorbed space for $(\underline{\underline{W}}_a, W+L)$. If (25.9) is satisfied then $\underline{\underline{F}}_a^{ref} = \underline{\underline{W}}([0,1])$ and the classification of extensions is the same as in Section 23 for bounded killing measure. If (25.9) fails then $\underline{\underline{F}}_a^{ref} = \underline{\underline{F}}$ and there are no nontrivial extensions with generator $\tilde{A}$ contained in $\mathcal{a}$. However $\underline{\underline{F}}_a^{env} = \underline{\underline{W}}([0,1])$ and the more general extension are classified by pairs $\mu_0(x)$, $\mu_1(x)$, again as in Section 23.

If $(\underline{\underline{F}}, E)$ is the absorbed space for $(\underline{\underline{L}}_a, L)$ and if (25.9) is satisfied then $\underline{\underline{F}}_a^{ref} = \underline{\underline{F}}$ and by the first remark at the end of Section 20, $(\underline{\underline{F}}, E)$ has no nontrivial extensions. If (25.9) fails then still $\underline{\underline{F}}_a^{ref} = \underline{\underline{F}}$ and $(\underline{\underline{F}}, E)$ cannot have extensions with generator $\tilde{A}$ contained in $\mathcal{a}$. However extensions do exist with $\Delta = \{0,1\}$. This is true whether or not the functions

$$\psi_0(x) = \mathcal{P}_x[X_{\zeta-0} = 0]$$

$$\psi_1(x) = \mathcal{P}_x[X_{\zeta-0} = 1]$$

are nontrivial. If ψ_0, ψ_1 are nontrivial then

$$\tilde{U}_{0,\infty}(\psi_0, \psi_0) = \tilde{U}_{0,\infty}(\psi_1, \psi_1) = +\infty$$

since ψ_0, ψ_1 are not in $\underline{\underline{F}}$ and so Theorem 20.2 -(i) is satisfied for any choice of μ_0, μ_1. If ψ_0, ψ_1 are trivial then $\kappa(x)\,dx$ has infinite energy and certainly μ_0, μ_1 can be chosen to satisfy Theorem 20.2-(i).

Remark. It is well known that ψ_0, ψ_1 are trivial for symmetric stable processes of index α, $0 < \alpha < 2$. We do not know if ψ_0, ψ_1 are ever nontrivial when $(\underline{\underline{F}}, E)$ is the absorbed space for $(\underline{\underline{L}}_a, L)$. ///

26. Stable Markov Chains

As in Section 17 let $\underset{=}{I}$ be a denumerably infinite set and let $m(x)$ be an everywhere positive function on $\underset{=}{I}$ which we view as a measure. Every symmetric submarkovian semigroup $\tilde{P}_t$ on $\tilde{L}(\underset{=}{I},m)$ corresponds to a unique standard transition matirx $\tilde{P}_t(x,y)$ satisfying the condition of symmetry (17.1) via

$$\tilde{P}_t f(x) = \Sigma_y \tilde{P}_t(x,y)f(y).$$

We will use both operator and matrix notation below.

Throughout this section $P(x,y)$ is an irreducible substochastic matrix on $\underset{=}{I}$ which vanishes on the diagonal and which is symmetrized by an everywhere positive measure α on I. (See (17.7).) Let $q, \mathcal{J}$, $\mathcal{E}$ and $\mathcal{J}^{\min}$ be as in Section 17 and define

$$\underset{=}{F} = \mathcal{J}^{\min} \cap L^2(m) \; ; \quad E = \mathcal{E} \; .$$

Clearly $(\underset{=}{F},E)$ is a regular Dirichlet space on $L^2(\underset{=}{I},m)$ and by Theorem 17.2 it corresponds to the well known minimal semigroup of Feller. It is also clear that

$$\underset{=a}{F}^{\text{ref}} = \mathcal{J} \cap L^2(m) \; .$$

Let Δ^{ref} be such that $\underset{=}{X}^{\text{ref}} = \underset{=}{I} \cup \Delta^{\text{ref}}$ is a regularizing space for $(\underset{=a}{F}^{\text{ref}}, E)$ and let ν and $(\underset{=}{N},N)$ be as in Section 20 for the pair $(\Delta^{\text{ref}},0)$. By Theorem 17.1 the semigroup $\tilde{P}_t$ has generator $\tilde{A}$ contained in the relevant local generator if and only if it satisfies the Kolmogorov equations (17.9) and (17.10). Now Theorems 20.2 and 20.5 together give

Theorem 26.1. *Let* $\tilde{P}_t(x,y)$ *be a standard transition matrix satisfying the condition of symmetry 17.2 and also the equivalent Kolmogorov equations (17.9) and (17.10). Then there exists a unique Dirichlet space* $(\underline{H},Q)$ *on* $L^2(\Delta^{ref},\nu)$ *satisfying Theorem 20.2-(iii) such that the associated Dirichlet space* $(\underline{F}^{\sim},E^{\sim})$ *is defined by Theorem 20.2-(iv) and the associated resolvent* $\tilde{G}_u, u > 0$ *is defined by (20.30). Conversely every Dirichlet space* $(\underline{H},Q)$ *on* $L^2(\Delta^{ref},\nu)$ *satisfying Theorem 20.2-(ii) determines in this manner a unique standard transition matrix* $\tilde{P}_t(x,y)$ *as above.* ///

Remark. This result was established in [47] for the special case when P is stochastic by entirely different techniques. One consequence of the approach there is that Δ^{ref} can be replaced by the active extremal Martin boundary. ///

For symmetric standard transition matrices which do not satisfy the appropriate Kolmogorov equations we have

Theorem 26.2. *Let* $\tilde{P}_t(x,y)$ *be a standard tranistion matrix satisfying the condition of symmetry 17.2. Then* $\tilde{P}_t(x,y)$ *satisfies* (17.2) *and* (17.5) *if and only if the associated Dirichlet space* $(\underline{F}^{\sim}, E^{\sim})$ *is an extension of* $(\underline{F},E)$.

Proof. If $(\underline{F}^{\sim}, E^{\sim})$ is an extension of (F,E) then by Lemma 1.1

(26.1) $$\operatorname{Lim}_{t\downarrow 0} (1/t)\{1-\tilde{P}_t(x,x)\}$$

$$= (1/m(x))\, E^{\sim}(e_x,e_x)$$

$$=(1/m(x))\, E(e_x,e_x)$$

$$= q(x)$$

$$\lim_{t \downarrow 0} (1/t)\, P^{\sim}_t(x,y)$$

$$= (1/m(x))\, E^{\sim}(e_x, e_y)$$

$$= (1/m(x))\, E(e_x, e_y)$$

$$= q(x)P(x,y)$$

and (17.5), (17.7) follow. Conversely let $P^{\sim}_t(x,y)$ satisfy 17.2, (17.2) and (17.5). By (17.2) the indicators e_x are all in $\underline{\underline{F}}^{\sim}$. By Lemma 1.1

$$\lim_{s \downarrow 0} (1/s)\{P^{\sim}_{t+s}(x,y) - P^{\sim}_t(x,y)\}$$

$$= E^{\sim}(e_x, P_t e_y)$$

and it follows that $P^{\sim}_t(x,y)$ is continuously differentiable for $t \geq 0$. Moreover from

$$P^{\sim}_{t+s}(x,y) - P^{\sim}_t(x,y) = \Sigma_{z \neq y} P^{\sim}_t(x,z)\, P^{\sim}_s(z,y)$$

$$+ P^{\sim}_t(x,y)\{P^{\sim}_s(y,y) - 1\}$$

and Fatou's lemma follows

$$(d/dt)\, P^{\sim}_t(x,y) \geq \Sigma_z\, P^{\sim}_t(x,z)q(z)P(z,y) - P^{\sim}_t(x,y)q(y).$$

(We refer to [5] where results of this sort are established without using symmetry.) Since $P^{min}_t(x,y)$ can be characterized as the minimal nonnegative solution of the Kolmogorov equations, it follows that $P^{\sim}_t$ dominates P^{min}_t and so by (20.43) $\underline{\underline{F}}_b$ is an ideal in $\underline{\underline{F}}^{\sim}_b$. That $E^{\sim}$ agrees with E on $\underline{\underline{F}}$ follows from (26.1) since the indicators e_x span $\underline{\underline{F}}$, and the theorem is proved. ///

Thus stable standard transition matrices satisfying a condition of symmetry can be classified as in Theorem 26.1 except that $(\Delta^{ref}, 0)$ must be replaced by a general pair (Δ, μ) satisfying condition (i) of Theorem 20.2. We will not state this formally as a theorem. By the first remakr at the end of Section 20 there is no loss of generality in taking $\Delta = \Delta^{ref}$ when

$$\sum_{x} \alpha(x) \{1 - P1(x)\} < +\infty . \tag{26.2}$$

In particular if (26.2) is satisfied and if $\underline{F} = \mathcal{J} \cap L^2(m)$ then the minimal transition matrix is the only one satisfying 17.2, (17.2) and (17.5). We do not know whether or not a "universal choice" for Δ is possible in general.

27. General Markov Chains

We continue to work with a standard transition $P_t(x,y)$ satisfying the condition of symmetry 17.2 but now we drop the assumption that every state is stable.

Let $(\underline{\underline{F}},E)$ be the associated Dirichlet space on $L^2(\underline{\underline{I}},m)$ and let $\mathcal{P}_x, x \in \underline{\underline{X}}$ and $X_t, t > 0$ be the associated Markov process with $\underline{\underline{X}}$ an appropriate regularizing space. For I a finite subset of $\underline{\underline{I}}$ let X_t^I be the time changed process of Section 8 with the role of ν played by the restriction of m to I. By the results in Section 8 the X_t^I form a Markov process which is symmetric with respect to m on I and the associated Dirichlet form E_I is given by

$$E_I(f,f) = E(H^I f, H^I f) \tag{27.1}$$

where H^I is the hitting operator for I. If is useful to note here that

$$E_I(f,f) = \inf \{E(g,g) : g \in \underline{\underline{F}}_{(e)} \text{ and } g = f \text{ on } I\} . \tag{27.2}$$

Also the Dirichlet norms $E_I(f,f)$ increase with I and

$$E(f,f) = \operatorname{Lim}_{I \uparrow \underline{\underline{I}}} E_I(f,f) \tag{27.3}$$

in the sense that f belong to $\underline{\underline{F}}_{(e)}$ if and only if the right side of (27.3) is finite and in this case (27.3) is valid. For each I there is a unique substochastic matrix P_I on I which vanishes on the diagonal and a unique measure α_I which symmetrizes P_I such that

$$(27.4)\qquad E_I(f,f) = \tfrac{1}{2}\Sigma\,\Sigma_{x,y\,\varepsilon\, I}\,\alpha_I(x)\,P_I(x,y)\{f(x)-f(y)\}^2$$

$$+\ \Sigma_{x\,\varepsilon\, I}\,\alpha_I(x)\{1-P_I 1(x)\}\,f^2(x)$$

and also

$$(27.5)\qquad \alpha_I(x) = m(x)\,q_I(x)$$

with q_I the rate for X_t^I. That is,

$$(27.5')\qquad q_I(x) = \mathrm{Lim}_{t\downarrow 0}\,(1/t)\{1-\theta_x(X_t^I = x)\}\,.$$

It is probabilistically obvious that if $I \subset J$ are finite, then P_I can be recovered from P_J via the formula

$$(27.6)\qquad P_I(x,y) = \Sigma_{k=0}^{\infty}\,\Sigma_{z\,\varepsilon\, K}(1_K P_J 1_K)^k(x,z)\,P_J(z,y)$$

where $x, y\,\varepsilon\, I$ and $K = (J-I)\cup\{x\}$. Moreover α_I is then determined by (27.4) and the fact that it must symmetrize P_I. Indeed it is not hard to show that actually

$$(27.7)\quad \alpha_I(x) = \alpha_J(x)\{1-\Sigma_{k=1}^{\infty}\Sigma_{z\,\varepsilon\, J-I}(P_J I_{J-1})^k(x,z)\,P_J(z,x)\}$$

(See Section 1 in [48].) Also it is obvious that irreducibility for $(\underline{\underline{F}}, E)$ guarantees that each P_I is irreducible and that $(\underline{\underline{F}}, E)$ is recurrent if and only if one and therefore every P_I is strictly stochastic.

A <u>road map</u> on $\underline{\underline{I}}$ is a family of matrices P_I with symmetrizing measures α_I indexed by I running over the finite subsets of $\underline{\underline{I}}$ such that (27.6) and (27.7) are satisfied when $I \subset J$. The road map will be called irreducible if the P_I are irreducible and recurrent if the P_I are strictly stochastic. Fix an irreducible road map and let the Dirichlet forms E_I be defined by (27.4). The following results are proved in Sections 1 and 2 of [48].

For general f on $\underline{\underline{I}}$ the Dirichlet norms $E_I(f,f)$ increase with I and therefore

(27.8) $$\mathcal{E}(f,f) = \operatorname{Lim}_{I \uparrow \underline{\underline{I}}} E_I(f,f)$$

is well defined. Let $\mathcal{F}$ be the totality of functions f on $\underline{\underline{I}}$ for which $\mathcal{E}(f,f) < +\infty$. We call $(\mathcal{F}, \mathcal{E})$ the discrete time Dirichlet space associated with the road map $\{P_I, \alpha_I\}$. For $f \in \mathcal{F}$

(27.9) $$E_I(f,f) = \inf \{\mathcal{E}(g,g) : g = f \text{ on } I\}$$

and indeed there exists a unique function denoted $H^I f$ such that this infimum is attained. That is, $H^I f = f$ on I and

(27.10) $$E_I(f,f) = \mathcal{E}(H^I f, H^I f)$$

Moreover H^I is the hitting operator for I relative to certain discrete time processes determined by the road map.

An everywhere positive measure m on $\underline{\underline{I}}$ is called a speed measure if

(27.11) $$\underline{\underline{F}} = \mathcal{F} \cap L^2(\underline{\underline{I}}, m)$$

is dense in $L^2(\underline{\underline{I}}, m)$. A speed measure m is called active if

(27.12) $$\underline{\underline{F}}_{(e)} = \mathcal{F}.$$

In general we can be sure that $\underline{\underline{F}}_{(e)}$ is contained in $\mathcal{F}$. It is clear from (27.10) that every bounded measure on $\underline{\underline{I}}$ is an active speed measure.

An unbounded measure may or may not be a speed measure and if it is a speed measure then it may or may not be active. Fix a speed measure m, let $P_t(x,y)$ be the standard transition matrix associated with the Dirichlet space $(\underline{\underline{F}},\mathcal{E})$ on $L^2(\underline{\underline{I}},m)$ and let $\{\tilde{P}_I, \tilde{\alpha}_I\}$ be the road map defined in the second paragraph. It follows from (27.2) and (27.9) on the one hand and from (27.3) and (27.8) on the other that $\{\tilde{P}_I, \tilde{\alpha}_I\}$ is identical with $\{P_I, \alpha_I\}$ if and only if m is active. We have proved

__Theorem 27.1.__ _Let $P_t(x,y)$ be a transition matrix satisfying the condition of symmetry 17.2 and let $(\underline{\underline{F}}, E)$ be the associated Dirichlet space on $L^2(I,m)$. For I a finite subset of $\underline{\underline{I}}$ let P_I, α_I be defined by (27.4) and (27.5). Then $\{P_I,\alpha_I\}$ is a road map and the extended Dirichlet space $(\underline{\underline{F}}_{(e)}, E)$ can be recovered via (27.3)._

Conversely let $\{P_I,\alpha_I\}$ be a road map with associated discrete time Dirichlet space $(\mathcal{F},\mathcal{E})$. Let m be a speed measure, let $\underline{\underline{F}} = \mathcal{F} \cap L^2(m)$ and let $P_t(x,y)$ be the standard transition matrix associated with the Dirichlet space $(\underline{\underline{F}}, E)$ on $L^2(\underline{\underline{I}},m)$. Then $\{P_I,\alpha_I\}$ is identical with the road map determined in the first paragraph if and only if m is active. ///

Theorem 27.1 is the main result in [48]. It can be viewed as extending Feller's notion of a road map and speed measure [15] to general symmetric Markov chains. We refer the reader also to D. Freedman's monograph [51] where the technique of time changing onto finite sets is applied to the study of nonsymmetric Markov chains with instantaneous states.

If $(\underline{\underline{F}},E)$ is a Dirichlet space on $L^2(\underline{\underline{I}},m)$ then the prescription (27.2), (27.4) can be applied directly to the spaces $(\underline{\underline{F}}^{ref},E)$, $(\underline{\underline{F}}^{res},E^{res})$ and $(\underline{\underline{F}}^{env},E^{res})$ to determine road maps. These will be identical with the ones determined by the corresponding active Dirichlet spaces in each case if and only if $\underline{\underline{F}}^{ref}$, $\underline{\underline{F}}^{res}$ or $\underline{\underline{F}}^{env}$ are the appropriate extended spaces.

To get an example of a speed measure which is not active let $\{P_I,\alpha_I\}$ be the road map determined by time changing reflected Brownian motion on $[0,1]$ with a bounded measure concentrated on and charging every point of a countable dense subset $\underline{\underline{I}}$ of $(0,1)$. Let m be a Radon measure on $(0,1)$ which is concentrated on $\underline{\underline{I}}$ and is unbounded near 0 and 1. Then m is a speed measure but is not active. The recovered road map corresponds to absorbing barrier Brownian motion.

Now let $\{P_I,\alpha_I\}$ be an irreducible generalized road map, let m be an active speed measure and let $(\underline{\underline{F}},E)$ be the associated Dirichlet space on $L^2(\underline{\underline{I}},m)$. In the remainder of this section we show that the road map determined by the enveloping space $(\underline{\underline{F}}^{env},E^{res})$ can be defined directly by suppressing incomplete excursions from a finite set I and then passing to the limit in I.

For each I define

$$E_I^o(f,f) = \tfrac{1}{2}\,\Sigma\Sigma_{x,y\,\varepsilon\, I}\,\alpha_I(x)P_I(x,y)\{f(x)-f(y)\}^2 .$$

Consider finite J containing I and let g on J agree with f on I. Let

$$P_J^o(x,y) = \begin{cases} P_J(x,y) & \text{for } x \neq y \\ 1 - P_J 1(x) & \text{for } x = y. \end{cases}$$

Clearly P_J^o is a possible road map for E_J^o with α_J the correct symmetrizing measure. (This makes sense even though P_J^o need not vanish on the diagonal.) Therefore either by Theorem 27.1 together with the results in Sections 7 and 8 or more directly by Sections 1 and 2 in [48]

$$E_J^o(g,g) \geq \tfrac{1}{2} \Sigma\Sigma_{x,y \varepsilon I}\, \alpha_J(x)\, Q^o(x,y)\, \{ f(x) - f(y) \}^2$$

$$E_I^o(f,f) = \tfrac{1}{2} \Sigma\Sigma_{x,y \varepsilon I}\, \alpha_J(x)\, Q(x,y)\, \{ f(x) - f(y) \}^2$$

where

$$Q^o(x,y) = P_J^o(x,y) + \Sigma_{z \varepsilon J-I} \Sigma_{k=1}^{\infty} (P_J^o 1_{J-I})^k (x,z)\, P_J^o(z,y)$$

$$Q(x,y) = P_J(x,y) + \Sigma_{z \varepsilon J-I} \Sigma_{k=1}^{\infty} (P_J 1_{J-I})^k (x,z)\, P(z,y).$$

But it is obvious that $Q^o(x,y) \geq Q(x,y)$ and we conclude that

$$E_J^o(g,g) \geq E_I^o(f,f). \tag{27.13}$$

On the other hand it is clear from (27.2) with J playing the role of $\underline{\underline{I}}$ that

$$\inf \{ E_J^o(g,g) : g = f \text{ on } I \} \leq E_I(f,f). \tag{27.14}$$

The inequalities (27.7) and (27.8) guarantee the existence of a limit

(27.15) $$E_I^{env}(f,f) = \mathrm{Lim}_{K \uparrow \underline{\underline{I}}} \inf\{ E_K^o(g,g) : g = f \text{ on } I \} .$$

Also if $J \supset I$ then

$$\inf\{ E_J^{env}(h,h) : h = f \text{ on } I\}$$

$$= \inf\{ \mathrm{Lim}_{K \uparrow I} \inf\{ E_K^o(g,g) : g = h \text{ on } J \} : h = f \text{ on } I \}$$

$$= \mathrm{Lim}_{K \uparrow I} \inf\{ E_K^o(g,g) : g = f \text{ on } I \}$$

$$= E_I^{env}(f,f).$$

In light of (27.9) this suggests the existence of a road map $\{ P_I^{env}, \alpha_I^{env} \}$ such that the E_I^{env} are the associated Dirichlet forms. Indeed the existence of $\{ P_I^{env}, \alpha_I^{env} \}$ follows immediately from Sections 1 and 2 in [48] and we take it for granted here. The result to be proved is

<u>Theorem 27.2</u>. <u>Let</u> $\{P_I, \alpha_I\}$ <u>be a road map and let</u> $\{ P_I^{env}, \alpha_I^{env} \}$ <u>be the road map corresponding to the Dirichlet forms</u> (27.15). <u>If</u> $\{ P_I, \alpha_I \}$ <u>corresponds to the Dirichlet space</u> $(\underline{\underline{F}}, E)$ <u>in the sense of Theorem 27.1, then the associated enveloping Dirichlet space</u> $(\underline{\underline{F}}^{env}, E^{res})$ <u>is identical with</u> $(\mathcal{F}^{env}, \mathcal{E}^{env})$, <u>the discrete time Dirichlet space for</u> $\{P_I^{env}, \alpha_I^{env}\}$. ///

There is no loss of generality in assuming that $\{ P_I, \alpha_I \}$ is irreducible and then the theorem is nontrivial only when $\{ P_I, \alpha_I \}$ and therefore $(\underline{\underline{F}}, E)$ is transient. For a proof in this case choose a regularizing space $\underline{\underline{X}}$ for $(\underline{\underline{F}}, E)$ and take for granted the notations of earlier sections. Fix D an open subset of $\underline{\underline{X}}$ with compact closure and let $f \in \underline{\underline{F}}_{loc}$ be bounded.

With the help of Lemma 14.2 it is easy to see that f has a representation

$$f = f^D + N^D(f\cdot\kappa) + H^M f$$

with f^D in the absorbed space $\underline{\underline{F}}^D$. Define

$$(27.16) \quad f^{env}(X_{\sigma(M)}) = I(\sigma(M) < +\infty)\, f(X_{\sigma(M)}) + I(\sigma(M) = +\infty;\ X_{\zeta-0} \in D) f(X_{\zeta-0})$$

$$f^{env}(X_{r(i)}) = I(r(i) < +\infty)\, f(X_{r(i)}) + I(r(i) = +\infty;\ X_{\zeta-0} \in D) f(X_{\zeta-0})$$

and then

$$(27.17) \quad M^{env} f(t) = Mf^D(t) + H^M f(X_t) + N^D f\cdot\kappa\,(X_t) \qquad \text{for} \quad X_t \in D$$

$$M^{env} f(\sigma(M)) = Mf^D(\sigma(M)) + f^{env}(X_{\sigma(M)})$$

$$M^{env} f(r(i)) = Mf^D(r(i)) + f^{env}(X_{r(i)}) .$$

On Ω and on Ω_∞ this functional is a martingale along excursions into D and the random measure $< M^{env} f >$ (dt) is well defined for the set of t such that $X_t \in D$. The proof of Lemma 20.7-(i) goes through unchanged with $\underline{\underline{X}}$ playing the role of Δ, $\mu(x, d\cdot) = \delta_x$ and with $\varphi = f$ to give

$$(27.18) \quad \tfrac{1}{2}\int_D < A_c f> (dx) + \tfrac{1}{2}\iint_{D\times D} J(dx, dw)\{f(x) - f(w)\}^2$$

$$+ \iint_{D\times M} J(dx, dy)\{f(x) - f(y)\}^2$$

$$= E(f^D, f^D) + \int_D \kappa(dx)\int_M H^M(x, dy)\{f(x) - f(y)\}^2$$

$$+ \tfrac{1}{2}\iint_{D\times D} (\kappa N^D \kappa)(dx, dw)\{f(x) - f(w)\}^2$$

$$+ \tfrac{1}{2}\,\delta\, \Sigma_i\, I(r(i) < \zeta)\, \{f(X_{r(i)}) - f(X_{e(i)-0})\}^2 .$$

If f belongs to $\underline{\underline{F}}_b^{env}$ then the left side of (27.18) remains bounded as $D \uparrow \underline{\underline{X}}$. Thus $E(f^D, f^D)$ is bounded independent of D and since $(\underline{\underline{F}}, E)$ is transient it follows that f has a representation

$$(27.19) \qquad f = f_o + N(f \cdot \varkappa) + h$$

with $f_o \in \underline{\underline{F}}_{(e)}$ and h harmonic and of course f_o, h are bounded. Let h be represented

$$H(x) = \mathscr{E}_x \Phi$$

with Φ a terminal variable as in Section 14. Define $M^{env}f(t)$ for all t by

$$(27.20) \qquad M^{env}f(t) = Mf_o(t) + h(X_t) + N(f \cdot \varkappa)(X_t) \quad \text{for } X_t \in \underline{\underline{X}}$$

$$M^{env}f(t) = Mf_o(t) + \Phi + I(X_{\zeta-0} \in \underline{\underline{X}})\, f(X_{\zeta-0}) \text{ for } t \geq \zeta$$

$$M^{env}f(t) = Mf_o(t) + \Phi \cdot \rho + I(X_{\zeta^*} \in \underline{\underline{X}})\, f(X_{\zeta^*}) \quad \text{for } t \leq \zeta^*.$$

This is consistent with (27.16) and from (6.10) follows with the help of Meyer's results on the decomposition of square integrable martingales

$$(27.21) \qquad E^{res}(f,f) = \tfrac{1}{2}\, \mathscr{E} \int_{\zeta^*}^{\zeta} \langle M^{env} f \rangle (dt).$$

Now fix I finite, denote the complement $\underline{\underline{X}} - I$ by I^c and let $\{e(i), r(i)\}$ be as in Section 13 with the open set I^c playing the role of D. If g belongs to $\underline{\underline{F}}$ then

$$\int \varkappa(dx) g(x) Ng \cdot \varkappa(x)$$

$$\leq \int \varkappa(dx) Ng^2 \cdot \varkappa(x)$$

$$\leq \int \varkappa(dy) g^2(y)$$

$$\leq E(g,g)$$

and therefore $Ng \cdot \varkappa \in \underline{\underline{F}}_{(e)}$. From this it follows that any f in $\underline{\underline{F}}^{env}$ can be approximated in the E^{res} sense by f such that in the representation (27.18)

(27.22) $$f_o = G\varphi$$

with φ bounded and integrable. Thus for the calculation which follows we can assume (27.22). We replace (27.16) by

(27.16') $$f^{env}(X_{\sigma(I)}) = I(\sigma(I) < +\infty) f(X_{\sigma(I)})$$

$$+ I(\sigma(I) = +\infty) \{ I(X_{\zeta-0} \in X) f(X_{\zeta-0}) + \Phi) \} .$$

etc.

and argue as in the proof of Lemma 20.7 to get

(27.23) $$\tfrac{1}{2} \mathscr{E} \int_{\zeta^*}^{\sigma(I)} < M^{env} f > (dt)$$

$$= \mathscr{E} \int_{\zeta^*}^{\sigma(I)} dt\, g(X_t)\, G^{I^c} g(X_t)$$

$$+ \tfrac{1}{2} \mathscr{E} \{ f^{env}(X_{\sigma(I)}) - f^{env}(X_{\zeta^*}) \}^2$$

$$+ \mathscr{E} \{ f^{env}(X_{\sigma(I)}) - f^{env}(X_{\zeta^*}) \} \int_{\zeta^*}^{\sigma(I)} dt\, g(X_t)$$

$$(27.24)\qquad \tfrac{1}{2}\mathcal{E}\,\Sigma_i\int_{e(i)}^{r(i)} <M^{env}f>(dt) + \tfrac{1}{2}\mathcal{E}\,\Sigma_i\,\{f(X_{e(i)})-f(X_{e(i)-0})\}^2$$

$$= \mathcal{E}\,\Sigma_i\int_{e(i)}^{r(i)} dt\, g(X_t)\, G^{I^c} g(X_t)$$

$$+ \tfrac{1}{2}\mathcal{E}\,\Sigma_i\,\{f^{env}(X_{r(i)}) - f(X_{e(i)-0})\}^2$$

$$+ \mathcal{E}\,\Sigma_i\,\{f^{env}(X_{r(i)}) - f(X_{e(i)-0})\}\int_{e(i)}^{r(i)} dt\, g(X_t).$$

As $I \uparrow \underline{\underline{I}}$ clearly (27.23) $\downarrow 0$ and therefore the left side of (27.24) has the same limit as

$$(25.25)\qquad \tfrac{1}{2}\mathcal{E}\int_{\zeta^*}^{\zeta} <M^{env}f>(dt)\, 1_{I^c}(dt) + \tfrac{1}{2}\int_I\int_{I^c} J(dx,dy)\,\{f(x)-f(y)\}^2.$$

Also

$$\mathcal{E}\,\Sigma_i\, I(r(i) < \zeta)\,\{f(X_{r(i)}) - f(X_{e(i)-0})\}\int_{e(i)}^{r(i)} dt\, g(X_t) = 0$$

and

$$\mathcal{E}\,\Sigma_i\, I(r(i) = \zeta)\,\{f^{env}(X_{r(i)}) - f(X_{e(i)-0})\}^2$$

$$\le \mathcal{E}\,\{f^{env}(X_{\sigma(I)}) - f^{env}(X_{\zeta^*})\}^2$$

which is dominated by (27.23) and therefore goes to zero and so the right side of (27.24) has the same limit as

$$(27.26)\qquad \tfrac{1}{2}\mathcal{E}\,\Sigma_i\, I(r(i) < \zeta)\,\{f(X_{r(i)}) - f(X_{e(i)-0})\}^2.$$

The proof of Lemma 21.1 guarantees

$$\tfrac{1}{2}\int_T < A^c f>(dx) = 0$$

and so $\frac{1}{2}\int\int_{I\times I} J(dx,dy)\{f(x)-f(y)\}^2$ can be added to the asymptotically equivalent expressions (25.25) and (27.26) to get

$$E^{res}(f,f)$$

$$= \operatorname{Lim}_{I\uparrow \underline{\underline{I}}} [\tfrac{1}{2}\mathscr{E}\,\Sigma_i I(r(i)<\zeta)\,\{f(X_{r(i)}) - f(X_{e(i)-0})\}^2$$

$$+ \tfrac{1}{2}\int\int_{I\times I} J(dx,dy)\,\{f(x) - f(y)\}^2\,]$$

$$= \operatorname{Lim}_{I\uparrow \underline{\underline{I}}} E^{o}_{I}(f,f).$$

By (27.13) and (27.17)

$$E^{o}_{I}(f,f) \le E^{env}_{I}(f,f) \le \operatorname{Lim}_{J\uparrow \underline{\underline{I}}} E^{o}_{J}(f,f).$$

Thus

$$(27.27)\qquad \mathscr{E}^{env}(f,f) = \operatorname{Lim}_{I\uparrow \underline{\underline{I}}} E^{o}_{I}(f,f)$$

and we have proved that $\underline{\underline{F}}^{env}$ is contained in $\mathscr{F}^{env}$ and $\mathscr{E}^{env} = E^{res}$ on $\underline{\underline{F}}^{env}$. To prove that actually, $\underline{\underline{F}}^{env} = \mathscr{F}^{env}$ fix open D with compact closure and let $(\underline{\underline{F}}^{(D)}, E^{(D)})$ be the time changed process of Section 8 with the role of ν played by the restriction of m to D. To estimate

$$E_I(f,f) \le E^{o}_{I}(f,f) + \Sigma_{x\in I}\,\alpha_I(x)\,\{1-P_I 1(x)\}\,g^2(x)$$

shows that $\mathscr{F}$ contains every f in $\mathscr{F}^{env}$ which is dominated by g in $\mathscr{F}$ and from this it follows that the restriction of $\mathscr{F}^{env}_b$ to D is contained in

$\underline{\underline{F}}^{(D)}$ and therefore in $\underline{\underline{F}}^{(D)\,env}$. If $\{P_I^{(D)}, \alpha_I^{(D)}\}$ is the road map determined by $(\underline{\underline{F}}^{(D)}, E^{(D)})$ then clearly $P_I^{(D)} = P_I$ and $\alpha_I^{(D)} = \alpha_I$ for I a finite subset of D and it follows from the first part of the proof that

$$E^{(D)\,res}(f,f) \leq \mathcal{E}^{env}(f,f).$$

Since the time changed semigroup $P_t^{(D)}$ dominates the absorbed semigroup for D the results in Section 21 guarantee that

$$E^{(D)res}(f,f) \geq \tfrac{1}{2} \iint_{D \times D} J(dx,dy)\{f(x) - f(y)\}^2$$

$$+ \tfrac{1}{2} \int_D \langle A_c f\rangle \,(dx)$$

and the theorem is proved.

Bibliography

1. A. Beurling and J. Deny, Dirichlet Spaces, Proc. Nat. Acad. Sci. U.S.A. 45 (1959), 208-215.

2. R. M. Blumenthal and R. K. Getoor, Markov Processes and Potential Theory, Academic Press, 1968.

3. H. Cartan, Sur les fondements de la théorie du potential, Bull. Soc. Math. France 691 (1941), 71-96.

4. __________, Théorie générale du balayage en potential newtonien, Ann. Univ. Grenoble, Sect. Sci. Math. Phys. (N.S) 22 (1946), 221-280.

5. K. L. Chung, Markov chains with Stationary Transition Probabilities, second edition, Springer-Verlag, 1967.

6. __________, Lectures on Boundary Theory for Markov Chains, Annals of Math. Studies No. 65, Princeton University Press, 1970.

7. __________, On the boundary theory for Markov chains, Acta. Math. 110 (1963), 19-77; Acta. Math. 115 (1966), 11-163.

8. J. Deny and J. L. Lions, Les espaces du type de Beppo Levi, Annales Inst. Fourier 5 (1953-4), 305-370.

9. J. L. Doob, Boundary properties of functions with finite Dirichlet integrals, Ann. Inst. Fourier 13 (1962), 573-621.

10. E. B. Dynkin, Theory of Markov Processes, Prentice Hall, Englewood Cliffs, 1961.

11. __________, Markov Processes I and II, Academic Press, New York, 1965.

12. J. Elliott, Dirichlet spaces associated with integro-differential operators I, II, Ill. Journ. of Math. 9(1965), 87-98; 10(1966), 66-89.

13. ________, Dirichlet spaces and boundary conditions for submarkovian resolvents, J. Math. Anal. App. 36 (1971), 251-282.

14. ________, On unsymmetric Dirichlet forms, Can. J. Math. 25(1973), 252 260.

15. W. Feller, On boundaries and lateral conditions for the Kolmogorov differential equations, Ann. of Math. 65 (1957), 527-570.

16. ________, An introduction to Probability Theory and its Applications, Vol. II, Wiley, New York, 1966.

17. S. R. Foguel, The Ergodic Theory of Markov Processes, Van Nostrand, 1969.

18. ________, On Feller's kernel and the Dirichlet norm, Nagoya Math. Journ. 24 (1964), 167-175.

19. M. Fukushima, A construction of reflecting barrier Brownian motions for bounded domains, Osaka J. Math. 4(1967), 183-215.

20. ________, On boundary conditions for multidimensional Brownian motion with symmetric resolvents, J. Math. Soc. Japan 21 (1969), 485-526.

21. ________, Regular representations of Dirichlet spaces, T.A.M.S. 155 (1971), 455-473.

22. ________, Dirichlet spaces and strong Markov processes, T.A.M.S. 162 (1971), 185-224.

23. ________, On transition probabilities of symmetric strong Markov processes, Journ. of Math. Kyota Univ. 12-3(1972), 431-450.

24. M. Fukushima, On the generation of Markov processes by symmetric forms, to appear in Proceedings of the second Japan-U.S.S.R. symposium on probability theory, Lecture Notes in Math., Springer-Verlag.

25. ______, Local property of Dirichlet forms and continuity of sample paths, to appear.

26. G. A. Hunt, Markov chains and Martin boundaries, Ill. J. Math. 4(1960), 313-340.

27. ______, Martingales et processus de Markov, Dunod, Paris, 1966.

28. K. Ito and H. P. McKean, Diffusion Processes and their Sample Paths, Academic Press, 1969.

29. J. G. Kemeny, A. W. Knapp and J. L. Snell, Denumerable Markov Chains, Van Nostrand, 1966.

30. H. Kunita, General boundary conditions for multi-dimensional diffusion processes, J. Math. Kyoto Univ. 10-2 (1970), 273-335.

31. ______, and S. Watanabe, On square integrable martingales, Nag. Math. Journ. 30 (1967), 209-245.

32. ______, and T. Watanabe, Some theorems concerning resolvents over locally compact spaces, Proc. Fifth Berk. Symp. on Math. Stat. and Prob. II, Part 2 (1967), 131-164.

33. M. Loeve, Probability Theory, Van Nostrand, 1955.

34. L. H. Loomis, An Introduction to Abstract Harmonic Analysis, Van Nostrand, 1953.

35. P. A. Meyer, Probability and Potentials, Blaisdell, 1966.

36. __________, Integrables stochastiques, Seminair de Probabilities vol. 1, No. 39, Springer Lecture Notes in Mathematics, 162-172.

37. __________, Processus de Markov: la frontiere de Martin, Lecture Notes in Mathematics No. 77, Springer Verlag, 1968.

38. L. Naïm, Surle role de la frontiere de R. S. Martin dans la theorie du potentiel, Ann. Inst. Fourier 7(1957), 183-281.

39. K. R. Parthasarathy, Probability Measures on Metric Spaces, Academic Press 1967.

40. F. Riesz and B. S. Nagy, Functional Analysis, Ungar, New York 1955.

41. G. C. Rota, An "Alternieven de Verfahren" for general positive operators, Bull. Amer. Math. Soc. 68 (1962), 95-102.

42. K. Sato and T. Ueno, Multidimensional diffusion processes and the Markov process on the boundary, J. Math. Kyoto Univ. 4(1965), 526-606.

43. T. Shiga and T. Watanabe, On Markov chains similar to the reflecting barrier Brownian motion, Osaka J. Math, 5(1968), 1-33.

44. M. L. Silverstein, Dirichlet spaces and random time change, Ill. J. Math., 17(1973), 1-72.

45. __________, Symmetric jump processes, to appear in the Trans. Amer. Math. Soc.

46. __________, The reflected Dirichlet space, to appear in the Ill. Journ.

47. __________, Classification of stable symmetric Markov chains, to appear in the Indiana Journal.

48. M. L. Silverstein, Symmetric Markov Chains, to appear in the Annals of Probability.

49. S. Watanabe, On discontinuous additive functionals and Lévy measures of a Markov process, Jap. J. Math. 34 (1966), 53-70.

50. M. Weil, Quasi-processus, Seminaire de Probabilities IV, Lecture Notes in Math. No. 124, Springer-Verlag, 1970.

51. D. Freedman, Approximating Countable Markov Chains, Holden Day, San Francisco, 1972.

52. H. Kesten, Hitting probabilities of single points for processes with stationary independent increments, Memoirs of the Amer. Math. Soc., 93 (1969).

53. W. Feller, Generalized second order differential equations and their lateral conditions, Ill. Journ. of Math. 1, 459-504 (1957).

54. J. Bretagnolle, Résultats de Kesten sur les processus à accroissements independants, Seminaire de Probabilities V, Lecture Notes in Mathematics 191, 21-36, Springer-Verlag (1971).

Comments added in proof

1. Theorem 4.9: For general Ray processes quasi-left-continuity is valid only when $\lim X_{T_n}$ is not a branching point. (See [32].) This complication never arises for symmetric processes because $G_1 C_{com}(\underline{\underline{X}})$ is quasi-uniformly dense in $C_{com}(\underline{\underline{X}})$.

2. Page 4.15: The first sentence is misleading. In general for nonsymmetric Markov processes there is a difference between accessible and predictable stopping times.

3. Top of page 5.3: The first p/q terms are treated separately because we are working with strong convergence in $\underline{\underline{F}}_{(e)}$.

4. (6.14): This estimate is never actually used.

5. Remark on page 9.4: Quasi-continuity of $\min(h, g_n)$ can be established directly.

6. (13.16): To guarantee that the right side converges we must assume also that $G\varphi$ is bounded.

7. Middle of page 15.4: Bounded harmonic functions belong to $\underline{\underline{F}}_{\ell oc}$. Also the approximation of $f^2 g$ by ψg depends on (15.6).

8. Theorem 18.3: This means $E^{\sim}(f, f) = E(f, f) - \int \nu(dx) f^2(x)$.

9. 20.3 Notation: The term $I(X^{\sim}_{\zeta-0} \varepsilon \Delta)\ \varphi(X^{\sim}_{\zeta-0})$ corresponds to the position of the $(\underline{\underline{F}}^{\sim}, E^{\sim})$ process after the $(\underline{\underline{F}}, E)$ process "drifts to ∂." The term $I(X^{\sim}_{\zeta-0} \varepsilon \underline{\underline{X}})\ \int \mu(X^{\sim}_{\zeta-0}, dy)\varphi(y)$ corresponds to the position after a "jump to ∂."

10. Middle of page 20.19: Regularity of (F, E) on $\underline{\underline{R}} - \{0\}$ is clear at least if $\kappa(x)$ increases monotonically as $x \downarrow 0$ and as $x \uparrow 0$.

11. (21.10): This should be written instead $(J^{\sim}-J)(A, X^{\sim}) \leq \kappa(A)$ since in general both sides of (21.10) are $+\infty$. The modified inequality can be established as follows. Let A be open and suppose $0 \leq f \leq 1$ and $f = 0$ on $\underline{\underline{X}} - A$. By (21.7) and (21.8)

$$\tfrac{1}{2} \iint J^{\sim}(dx, dw) \{f(x) - f(w)\}^2$$

$$\leq \int \kappa(dx) f^2(x) + \tfrac{1}{2} \iint J(dx, dw) \{f(x)-f(w)\}^2$$

and by (21.5)

$$\tfrac{1}{2} \iint_{A \times A} J^{\sim}(dx, dw) \{f(x)-f(w)\}^2$$

$$\geq \tfrac{1}{2} \iint_{A \times A} J(dx, dw) \{f(x) - f(w)\}^2$$

and so

$$\int J^{\sim}(dx, \underline{\underline{X}}^{\sim} - A) f^2(x) \leq \int \kappa(dx) f^2(x)$$

$$+ \int J(dx, \underline{\underline{X}} - A) f^2(x).$$

The desired inequality follows upon letting f increase to the indicator 1_A and then varying A.

12. Line 7 on page 25.3: The condition $fg = 0$ is redundant.

13. Bottom of page 25.5 and page 25.6. In fact ψ_0, ψ_1 are always trivial for $(\underline{\underline{L}}_a, L)$ and are never trivial for $(\underline{\underline{W}}_a, W+L)$. This is completely proved in a paper of P. W. Millar, "Exit properties of stochastic processes with stationary independent increments", TAMS 178(1973), 459-477. Partial results are contained in a paper by N. Ikeda and S. Watanabe, "The local structure of a class of diffusions and related

problems" which appears in Proceedings of the Second Japan-USSR Symposium on Probability Theory, Springer Verlay Lecture Notes in Mathematics No. 330.

14. (27.12): This means that $\mathcal{J} \cap L^2(m)$ is $\mathcal{E}$ dense in $\mathcal{J}$ in the sense of paragraph 1.6.

15. Middle of page 27.11: We are also using $\mathcal{E}\int_{\zeta_*}^{\zeta} dt\, g(X_t) G^{I_c} g(X_t) \to 0$ to conclude that the right side of (27.24) is asymptotically equivalent to (27.26).

Index

Page	Misprints and minor corrections
1.1.	line 3: delete final period
1.2.	(1.4): $E_u(f,f)$
1.11.	line 5 from b: in
1.14.	line 4 - line 6: replace 1,E by $1^{\frac{1}{2}}, E^{\frac{1}{2}}$
2.4.	(2.1): $\lvert f_i(x)-\gamma(f_i)\rvert$
3.1.	delete last sentence in first paragraph
3.3.	line 3, line 7: $\underline{\underline{F}}$
3.6.	line 9: delete final period
3.7.	line 7: insert "on G_m"
3.9.	(3.9): $\lvert f(x)\rvert$
3.11.	line 12: $\int_{U-\omega} dx$
3.15.	line 8: insert "Here $U = \bigcup_n U_n$". after period
4.2.	line 6 from b: adjoin
4.7.	include first sentence of Corollary 4.7 under (i)
4.13.	line 3: X_{T_n}
	line 6: P_{t_m}
4.17.	line 8 from b: replace $\underline{\underline{X}}$ by G
4.18.	line 1: delete superscript *
	line 8: insert "with compact closure" after G_n
6.8.	(6.15): $3\ E\{\int_{\zeta^*}^{\zeta} ds\lvert\varphi(X_s)\rvert\}^2$
8.2.	line 1 from b: $(\beta-\alpha)\int_0^{R_u} a(\nu;ds)e^{-\beta a(\nu;s)}$ etc.
8.8.	line 4, line 5, line 9: delete superscript M
	line 8: replace superscript M by ν
8.9.	line 5: replace dx by $\nu(dx)$
8.12.	line 15: replace "Lemma 8.7" by "Theorem 8.5"
9.1.	line 2 from b: replace 2.15 by 3.15
9.2.	line 10: replace 9.2.2 by 9.1.2
9.4.	line 1: shows
10.4.	line 1: nonnegative

10.7.	line 4 from b: replace "$\sigma(M)<\zeta$" by "$\sigma(M')<\zeta$"
11.1.	line 11: replace "$t\uparrow\infty$" by "$t\downarrow 0$"
11.2.	line 2: replace X_t by X_s
11.3.	line 3: replace quasi- by almost
11.5.	line 9: replace $\geq$ by $\leq$
11.7.	line 6: replace Mf(x) by Mf(s)
	line 3 from b: replace $\langle A_t f'\rangle$ by $\langle A_t f\rangle - \langle A_t f'\rangle$
11.9.	line 9: replace "if" by "it"
11.10.	line 9: replace σ by τ (in three places)
11.11.	(11.30): $\langle A_c f'\rangle$ (dx)
11.12.	line 3 from b: insert "$I(\tau<\infty)$" inside expectation
III.1.	line 4: Chapter II
III.2.	line 9: coincides
13.2.	(13.5) replace "Φ" by "$\Phi I(\sigma(M)=+\infty)$"
13.8.	delete line 10
14.3.	line 13: 14.1
15.2.	line 10: $H_u\, G_u^{\sim}\varphi$
16.3.	line 3 from b: whenever
16.4.	line 2 from b: (16.2)
16.5.	line 5: $f_\circ^2(X_R)$
16.6.	line 9: Lemma 14.2
	line 13: $f = f_\circ + h$ with $f_\circ$
17.2.	line 7: instantaneous
17.3.	line 2: irreducibility
19.1.	line 9: replace $\underline{\underline{F}}_0$ by $\underline{\underline{F}}^o$ and $E(f,f)$ by $E^o(f,f)$
19.2.	line 11: $q' : \underline{\underline{X}}'\cup\{\partial'\} \to \underline{\underline{X}}\cup\{\partial\}$
20.3.	line 2 from b: $\underline{\underline{H}}^{\sim}$
20.4.	line 2: $(\underline{\underline{F}},E)$
20.7.	(20.25) $X^{\sim}_{\zeta-0}$
	line 6, line 7 from b: $p^{\sim}(X_t)$

	line 4 from b: insert "$I(X^{\sim}_{\zeta*} = \partial^{\sim})$" inside E
	line 3 from b: insert "$I(X^{\sim}_{\zeta*} = \partial^{\sim})$" inside E
20.11.	line 5: delete $\frac{1}{2}$
20.12.	line 6: $H^{\sim}(X_t, dy)$
20.14.	line 8: G_V
20.15.	line 2: $\frac{1}{2}u\int dx$
20.16.	line 3: $\frac{1}{2}U^{\sim}_{0,\infty}$
	line 2 from b: $G_1 g(x)$
20.19.	line 6 from b: spaces
20.25.	line 1: functionals
	line 3, line 4 from b: $f^{\sim}$
20.29.	line 2: insert "$-f(X_0)$" inside $\{\ \}$
20.33.	line 4: replace (20.65) by (20.67)
	line 7: replace (20.57), (20.67), (20.67') by (20.67). (20.64), (20.64')
21.1.	line 13: representation
23.6.	line 2 from b: $\int$(23.16)
23.8.	(23.19): $\cdots -\int \varkappa(dx)\ (r(x)+\mu_0(x))\psi_1(x)$
23.9.	line 2 from b: $\psi^o_0(x)$
23.12.	line 7: $\psi^o_0(x)$
24.4.	line 8: $\{\gamma f(\infty) - f(0)\}$
25.1.	line 3 from b: $\underline{\underline{W}}_a$
25.2.	line 8: replace $2 + 2(x)^2$ by $2 + 2\|x\|$
25.3.	line 6: $\int dx\ 1_A(x)\ f(x) \int \ell(dy)\ 1_{A^c}(x+y)\ g(x+y) = 0$
25.5.	line 9: extensions
26.3.	line 5: replace (17.7) by (17.2)
26.4.	line 4: remark
27.1.	line 13: It
27.6.	line 1 from b: (27.13) and (27.14)
27.9.	line 7: $h(x)$
27.10.	line 6: (27.19)

(27.22): Gg

line 8: replace φ by g

(27.16') delete last)

27.11. line 7: (27.25)/ $1_{I^c}(t)$ / $\iint_{I\times I^c}$

27.12. line 2: replace (25.25) by (27.25)

line 7: replace (27.17) by (27.15)

line 4 from b: replace "To" by "The"

Vol. 247: Lectures on Operator Algebras. Tulane University Ring and Operator Theory Year, 1970–1971. Volume II. XI, 786 pages. 1972. DM 40,–

Vol. 248: Lectures on the Applications of Sheaves to Ring Theory. Tulane University Ring and Operator Theory Year, 1970–1971. Volume III. VIII, 315 pages. 1971. DM 26,–

Vol. 249: Symposium on Algebraic Topology. Edited by P. J. Hilton. VII, 111 pages. 1971. DM 16,–

Vol. 250: B. Jónsson, Topics in Universal Algebra. VI, 220 pages. 1972. DM 20,–

Vol. 251: The Theory of Arithmetic Functions. Edited by A. A. Gioia and D. L. Goldsmith VI, 287 pages. 1972. DM 24,–

Vol. 252: D. A. Stone, Stratified Polyhedra. IX, 193 pages. 1972. DM 18,–

Vol. 253: V. Komkov, Optimal Control Theory for the Damping of Vibrations of Simple Elastic Systems. V, 240 pages. 1972. DM 20,–

Vol. 254: C. U. Jensen, Les Foncteurs Dérivés de $\varprojlim$ et leurs Applications en Théorie des Modules. V, 103 pages. 1972. DM 16,–

Vol. 255: Conference in Mathematical Logic – London '70. Edited by W. Hodges. VIII, 351 pages. 1972. DM 26,–

Vol. 256: C. A. Berenstein and M. A. Dostal, Analytically Uniform Spaces and their Applications to Convolution Equations. VII, 130 pages. 1972. DM 16,–

Vol. 257: R. B. Holmes, A Course on Optimization and Best Approximation. VIII, 233 pages. 1972. DM 20,–

Vol. 258: Séminaire de Probabilités VI. Edited by P. A. Meyer. VI, 253 pages. 1972. DM 22,–

Vol. 259: N. Moulis, Structures de Fredholm sur les Variétés Hilbertiennes. V, 123 pages. 1972. DM 16,–

Vol. 260: R. Godement and H. Jacquet, Zeta Functions of Simple Algebras. IX, 188 pages. 1972. DM 18,–

Vol. 261: A. Guichardet, Symmetric Hilbert Spaces and Related Topics. V, 197 pages. 1972. DM 18,–

Vol. 262: H. G. Zimmer, Computational Problems, Methods, and Results in Algebraic Number Theory. V, 103 pages. 1972. DM 16,–

Vol. 263: T. Parthasarathy, Selection Theorems and their Applications. VII, 101 pages. 1972. DM 16,–

Vol. 264: W. Messing, The Crystals Associated to Barsotti-Tate Groups: With Applications to Abelian Schemes. III, 190 pages. 1972. DM 18,–

Vol. 265: N. Saavedra Rivano, Catégories Tannakiennes. II, 418 pages. 1972. DM 26,–

Vol. 266: Conference on Harmonic Analysis. Edited by D. Gulick and R. L. Lipsman. VI, 323 pages. 1972. DM 24,–

Vol. 267: Numerische Lösung nichtlinearer partieller Differential- und Integro-Differentialgleichungen. Herausgegeben von R. Ansorge und W. Törnig, VI, 339 Seiten. 1972. DM 26,–

Vol. 268: C. G. Simader, On Dirichlet's Boundary Value Problem. IV, 238 pages. 1972. DM 20,–

Vol. 269: Théorie des Topos et Cohomologie Etale des Schémas. (SGA 4). Dirigé par M. Artin, A. Grothendieck et J. L. Verdier. XIX, 525 pages. 1972. DM 50,–

Vol. 270: Théorie des Topos et Cohomologie Etale des Schémas. Tome 2. (SGA 4). Dirigé par M. Artin, A. Grothendieck et J. L. Verdier. V, 418 pages. 1972. DM 50,–

Vol. 271: J. P. May, The Geometry of Iterated Loop Spaces. IX, 175 pages. 1972. DM 18,–

Vol. 272: K. R. Parthasarathy and K. Schmidt, Positive Definite Kernels, Continuous Tensor Products, and Central Limit Theorems of Probability Theory. VI, 107 pages. 1972. DM 16,–

Vol. 273: U. Seip, Kompakt erzeugte Vektorräume und Analysis. IX, 119 Seiten. 1972. DM 16,–

Vol. 274: Toposes, Algebraic Geometry and Logic. Edited by. F. W. Lawvere. VI, 189 pages. 1972. DM 18,–

Vol. 275: Séminaire Pierre Lelong (Analyse) Année 1970–1971. VI, 181 pages. 1972. DM 18,–

Vol. 276: A. Borel, Représentations de Groupes Localement Compacts. V, 98 pages. 1972. DM 16,–

Vol. 277: Séminaire Banach. Edité par C. Houzel. VII, 229 pages. 1972. DM 20,–

Vol. 278: H. Jacquet, Automorphic Forms on GL(2). Part II. XIII, 142 pages. 1972. DM 16,–

Vol. 279: R. Bott, S. Gitler and I. M. James, Lectures on Algebraic and Differential Topology. V, 174 pages. 1972. DM 18,–

Vol. 280: Conference on the Theory of Ordinary and Partial Differential Equations. Edited by W. N. Everitt and B. D. Sleeman. XV, 367 pages. 1972. DM 26,–

Vol. 281: Coherence in Categories. Edited by S. Mac Lane. VII, 235 pages. 1972. DM 20,–

Vol. 282: W. Klingenberg und P. Flaschel, Riemannsche Hilbertmannigfaltigkeiten. Periodische Geodätische. VII, 211 Seiten. 1972. DM 20,–

Vol. 283: L. Illusie, Complexe Cotangent et Déformations II. VII, 304 pages. 1972. DM 24,–

Vol. 284: P. A. Meyer, Martingales and Stochastic Integrals I. VI, 89 pages. 1972. DM 16,–

Vol. 285: P. de la Harpe, Classical Banach-Lie Algebras and Banach-Lie Groups of Operators in Hilbert Space. III, 160 pages. 1972. DM 16,–

Vol. 286: S. Murakami, On Automorphisms of Siegel Domains. V, 95 pages. 1972. DM 16,–

Vol. 287: Hyperfunctions and Pseudo-Differential Equations. Edited by H. Komatsu. VII, 529 pages. 1973. DM 36,–

Vol. 288: Groupes de Monodromie en Géométrie Algébrique. (SGA 7 I). Dirigé par A. Grothendieck. IX, 523 pages. 1972. DM 50,–

Vol. 289: B. Fuglede, Finely Harmonic Functions. III, 188. 1972. DM 18,–

Vol. 290: D. B. Zagier, Equivariant Pontrjagin Classes and Applications to Orbit Spaces. IX, 130 pages. 1972. DM 16,–

Vol. 291: P. Orlik, Seifert Manifolds. VIII, 155 pages. 1972. DM 16,–

Vol. 292: W. D. Wallis, A. P. Street and J. S. Wallis, Combinatorics: Room Squares, Sum-Free Sets, Hadamard Matrices. V, 508 pages. 1972. DM 50,–

Vol. 293: R. A. DeVore, The Approximation of Continuous Functions by Positive Linear Operators. VIII, 289 pages. 1972. DM 24,–

Vol. 294: Stability of Stochastic Dynamical Systems. Edited by R. F. Curtain. IX, 332 pages. 1972. DM 26,–

Vol. 295: C. Dellacherie, Ensembles Analytiques, Capacités, Mesures de Hausdorff. XII, 123 pages. 1972. DM 16,–

Vol. 296: Probability and Information Theory II. Edited by M. Behara, K. Krickeberg and J. Wolfowitz. V, 223 pages. 1973. DM 20,–

Vol. 297: J. Garnett, Analytic Capacity and Measure. IV, 138 pages. 1972. DM 16,–

Vol. 298: Proceedings of the Second Conference on Compact Transformation Groups. Part 1. XIII, 453 pages. 1972. DM 32,–

Vol. 299: Proceedings of the Second Conference on Compact Transformation Groups. Part 2. XIV, 327 pages. 1972. DM 26,–

Vol. 300: P. Eymard, Moyennes Invariantes et Représentations Unitaires. II, 113 pages. 1972. DM 16,–

Vol. 301: F. Pittnauer, Vorlesungen über asymptotische Reihen. VI, 186 Seiten. 1972. DM 18,–

Vol. 302: M. Demazure, Lectures on p-Divisible Groups. V, 98 pages. 1972. DM 16,–

Vol. 303: Graph Theory and Applications. Edited by Y. Alavi, D. R. Lick and A. T. White. IX, 329 pages. 1972. DM 26,–

Vol. 304: A. K. Bousfield and D. M. Kan, Homotopy Limits, Completions and Localizations. V, 348 pages. 1972. DM 26,–

Vol. 305: Théorie des Topos et Cohomologie Etale des Schémas. Tome 3. (SGA 4). Dirigé par M. Artin, A. Grothendieck et J. L. Verdier. VI, 640 pages. 1973. DM 50,–

Vol. 306: H. Luckhardt, Extensional Gödel Functional Interpretation. VI, 161 pages. 1973. DM 18,–

Vol. 307: J. L. Bretagnolle, S. D. Chatterji et P.-A. Meyer, Ecole d'été de Probabilités: Processus Stochastiques. VI, 198 pages. 1973. DM 20,–

Vol. 308: D. Knutson, λ-Rings and the Representation Theory of the Symmetric Group. IV, 203 pages. 1973. DM 20,–

Vol. 309: D. H. Sattinger, Topics in Stability and Bifurcation Theory. VI, 190 pages. 1973. DM 18,–

Vol. 310: B. Iversen, Generic Local Structure of the Morphisms in Commutative Algebra. IV, 108 pages. 1973. DM 16,-

Vol. 311: Conference on Commutative Algebra. Edited by J. W. Brewer and E. A. Rutter. VII, 251 pages. 1973. DM 22,-

Vol. 312: Symposium on Ordinary Differential Equations. Edited by W. A. Harris, Jr. and Y. Sibuya. VIII, 204 pages. 1973. DM 22,-

Vol. 313: K. Jörgens and J. Weidmann, Spectral Properties of Hamiltonian Operators. III, 140 pages. 1973. DM 16,-

Vol. 314: M. Deuring, Lectures on the Theory of Algebraic Functions of One Variable. VI, 151 pages. 1973. DM 16,-

Vol. 315: K. Bichteler, Integration Theory (with Special Attention to Vector Measures). VI, 357 pages. 1973. DM 26,-

Vol. 316: Symposium on Non-Well-Posed Problems and Logarithmic Convexity. Edited by R. J. Knops. V, 176 pages. 1973. DM 18,-

Vol. 317: Séminaire Bourbaki – vol. 1971/72. Exposés 400–417. IV, 361 pages. 1973. DM 26,-

Vol. 318: Recent Advances in Topological Dynamics. Edited by A. Beck, VIII, 285 pages. 1973. DM 24,-

Vol. 319: Conference on Group Theory. Edited by R. W. Gatterdam and K. W. Weston. V, 188 pages. 1973. DM 18,-

Vol. 320: Modular Functions of One Variable I. Edited by W. Kuyk. V, 195 pages. 1973. DM 18,-

Vol. 321: Séminaire de Probabilités VII. Edité par P. A. Meyer. VI, 322 pages. 1973. DM 26,-

Vol. 322: Nonlinear Problems in the Physical Sciences and Biology. Edited by I. Stakgold, D. D. Joseph and D. H. Sattinger. VIII, 357 pages. 1973. DM 26,-

Vol. 323: J. L. Lions, Perturbations Singulières dans les Problèmes aux Limites et en Contrôle Optimal. XII, 645 pages. 1973. DM 42,-

Vol. 324: K. Kreith, Oscillation Theory. VI, 109 pages. 1973. DM 16,-

Vol. 325: Ch.-Ch. Chou, La Transformation de Fourier Complexe et L'Equation de Convolution. IX, 137 pages. 1973. DM 16,-

Vol. 326: A. Robert, Elliptic Curves. VIII, 264 pages. 1973. DM 22,-

Vol. 327: E. Matlis, 1-Dimensional Cohen-Macaulay Rings. XII, 157 pages. 1973. DM 18,-

Vol. 328: J. R. Büchi and D. Siefkes, The Monadic Second Order Theory of All Countable Ordinals. VI, 217 pages. 1973. DM 20,-

Vol. 329: W. Trebels, Multipliers for (C, α)-Bounded Fourier Expansions in Banach Spaces and Approximation Theory. VII, 103 pages. 1973. DM 16,-

Vol. 330: Proceedings of the Second Japan-USSR Symposium on Probability Theory. Edited by G. Maruyama and Yu. V. Prokhorov. VI, 550 pages. 1973. DM 36,-

Vol. 331: Summer School on Topological Vector Spaces. Edited by L. Waelbroeck. VI, 226 pages. 1973. DM 20,-

Vol. 332: Séminaire Pierre Lelong (Analyse) Année 1971-1972. V, 131 pages. 1973. DM 16,-

Vol. 333: Numerische, insbesondere approximationstheoretische Behandlung von Funktionalgleichungen. Herausgegeben von R. Ansorge und W. Törnig. VI, 296 Seiten. 1973. DM 24,-

Vol. 334: F. Schweiger, The Metrical Theory of Jacobi-Perron Algorithm. V, 111 pages. 1973. DM 16,-

Vol. 335: H. Huck, R. Roitzsch, U. Simon, W. Vortisch, R. Walden, B. Wegner und W. Wendland, Beweismethoden der Differentialgeometrie im Großen. IX, 159 Seiten. 1973. DM 18,-

Vol. 336: L'Analyse Harmonique dans le Domaine Complexe. Edité par E. J. Akutowicz. VIII, 169 pages. 1973. DM 18,-

Vol. 337: Cambridge Summer School in Mathematical Logic. Edited by A. R. D. Mathias and H. Rogers. IX, 660 pages. 1973. DM 42,-

Vol: 338: J. Lindenstrauss and L. Tzafriri, Classical Banach Spaces. IX, 243 pages. 1973. DM 22,-

Vol. 339: G. Kempf, F. Knudsen, D. Mumford and B. Saint-Donat, Toroidal Embeddings I. VIII, 209 pages. 1973. DM 20,-

Vol. 340: Groupes de Monodromie en Géométrie Algébrique. (SGA 7 II). Par P. Deligne et N. Katz. X, 438 pages. 1973. DM 40,-

Vol. 341: Algebraic K-Theory I, Higher K-Theories. Edited by H. Bass. XV, 335 pages. 1973. DM 26,-

Vol. 342: Algebraic K-Theory II, "Classical" Algebraic K-Theory, and Connections with Arithmetic. Edited by H. Bass. XV, 527 pages. 1973. DM 36,-

Vol. 343: Algebraic K-Theory III, Hermitian K-Theory and Geometric Applications. Edited by H. Bass. XV, 572 pages. 1973. DM 38,-

Vol. 344: A. S. Troelstra (Editor), Metamathematical Investigation of Intuitionistic Arithmetic and Analysis. XVII, 485 pages. 1973. DM 34,-

Vol. 345: Proceedings of a Conference on Operator Theory. Edited by P. A. Fillmore. VI, 228 pages. 1973. DM 20,-

Vol. 346: Fučik et al., Spectral Analysis of Nonlinear Operators. II, 287 pages. 1973. DM 26,-

Vol. 347: J. M. Boardman and R. M. Vogt, Homotopy Invariant Algebraic Structures on Topological Spaces. X, 257 pages. 1973. DM 22,-

Vol. 348: A. M. Mathai and R. K. Saxena, Generalized Hypergeometric Functions with Applications in Statistics and Physical Sciences. VII, 314 pages. 1973. DM 26,-

Vol. 349: Modular Functions of One Variable II. Edited by W. Kuyk and P. Deligne. V, 598 pages. 1973. DM 38,-

Vol. 350: Modular Functions of One Variable III. Edited by W. Kuyk and J.-P. Serre. V, 350 pages. 1973. DM 26,-

Vol. 351: H. Tachikawa, Quasi-Frobenius Rings and Generalizations. XI, 172 pages. 1973. DM 18,-

Vol. 352: J. D. Fay, Theta Functions on Riemann Surfaces. V, 137 pages. 1973. DM 16,-

Vol. 353: Proceedings of the Conference. on Orders, Group Rings and Related Topics. Organized by J. S. Hsia, M. L. Madan and T. G. Ralley. X, 224 pages. 1973. DM 20,-

Vol. 354: K. J. Devlin, Aspects of Constructibility. XII, 240 pages. 1973. DM 22,-

Vol. 355: M. Sion, A Theory of Semigroup Valued Measures. V, 140 pages. 1973. DM 16,-

Vol. 356: W. L. J. van der Kallen, Infinitesimally Central-Extensions of Chevalley Groups. VII, 147 pages. 1973. DM 16,-

Vol. 357: W. Borho, P. Gabriel und R. Rentschler, Primideale in Einhüllenden auflösbarer Lie-Algebren. V, 182 Seiten. 1973. DM 18,-

Vol. 358: F. L. Williams, Tensor Products of Principal Series Representations. VI, 132 pages. 1973. DM 16,-

Vol. 359: U. Stammbach, Homology in Group Theory. VIII, 183 pages. 1973. DM 18,-

Vol. 360: W. J. Padgett and R. L. Taylor, Laws of Large Numbers for Normed Linear Spaces and Certain Fréchet Spaces. VI, 111 pages. 1973. DM 16,-

Vol. 361: J. W. Schutz, Foundations of Special Relativity: Kinematic Axioms for Minkowski Space Time. XX, 314 pages. 1973. DM 26,-

Vol. 362: Proceedings of the Conference on Numerical Solution of Ordinary Differential Equations. Edited by D. Bettis. VIII, 490 pages. 1974. DM 34,-

Vol. 363: Conference on the Numerical Solution of Differential Equations. Edited by G. A. Watson. IX, 221 pages. 1974. DM 20,-

Vol. 364: Proceedings on Infinite Dimensional Holomorphy. Edited by T. L. Hayden and T. J. Suffridge. VII, 212 pages. 1974. DM 20,-

Vol. 365: R. P. Gilbert, Constructive Methods for Elliptic Equations. VII, 397 pages. 1974. DM 26,-

Vol. 366: R. Steinberg, Conjugacy Classes in Algebraic Groups (Notes by V. V. Deodhar). VI, 159 pages. 1974. DM 18,-

Vol. 367: K. Langmann und W. Lütkebohmert, Cousinverteilungen und Fortsetzungssätze. VI, 151 Seiten. 1974. DM 16,-

Vol. 368: R. J. Milgram, Unstable Homotopy from the Stable Point of View. V, 109 pages. 1974. DM 16,-

Vol. 369: Victoria Symposium on Nonstandard Analysis. Edited by A. Hurd and P. Loeb. XVIII, 339 pages. 1974. DM 26,-

Vol. 370: B. Mazur and W. Messing, Universal Extensions and One Dimensional Crystalline Cohomology. VII, 134 pages. 1974. DM 16,-